HOME
WATERS

HOME WATERS

DISCOVERING THE SUBMERGED
SCIENCE OF BRITAIN'S COAST

DAVID GEORGE BOWERS

**ADLARD
COLES**

LONDON · OXFORD · NEW YORK · NEW DELHI · SYDNEY

ADLARD COLES
Bloomsbury Publishing Plc
50 Bedford Square, London, WC1B 3DP, UK
29 Earlsfort Terrace, Dublin 2, Ireland

BLOOMSBURY, ADLARD COLES and the Adlard Coles logo are trademarks
of Bloomsbury Publishing Plc

First published in Great Britain 2023

A catalogue record for this book is available from the British Library

Library of Congress Cataloguing-in-Publication data has been applied for

ISBN: PB: 978-1-4729-9068-6; ePub: 978-1-4729-9069-3;
ePDF: 978-1-4729-9067-9

2 4 6 8 10 9 7 5 3 1

Typeset in Minion Pro by Deanta Global Publishing Services, Chennai, India
Printed and bound in Great Britain by CPI Trade

To find out more about our authors and books visit www.bloomsbury.com
and sign up for our newsletters

CONTENTS

To
Amy, Alice, Anwen and Zoe,
of whom I am so proud

PREFACE

ON A TRIP TO FOYLE'S BOOKSHOP in Charing Cross Road in London, I bought two books about the sea and took them to read in a nearby coffee shop. One book was disappointing and I soon put it down, but the other had me turning the pages while my coffee grew cold. It was called *How the Ocean Works* and was written by Mark Denny. The book described the ways that the physics, chemistry and biology of the ocean combine to allow life to exist and thrive in an environment that is very different to the one we are used to on land.

How the Ocean Works – and indeed most books about the science of the sea – is about the big, deep ocean. It's easy to see why. Nearly all the salt water on our planet is found in the major ocean basins – the Pacific, Atlantic and Indian. From the point of view of the Earth as a system, it is these deep oceans that matter most. But from a *people* perspective, things are different. For most of us, the nearest stretch of salt water lies on the continental shelf – the edge of the continents, flooded when sea level rose at the end of the last ice age. These seas, called shelf seas, behave differently to the deep ocean; they are shallow, tides are important and fresh water flows in from the land.

There are very few books that deal with the science of the shelf seas in a way that is accessible to an interested but non-specialist reader. This is the sort of book that I wanted to write. You will find an account of the science of shelf seas here, but it is mixed in with a bit of history, some travel to places of significance to the stories I want to tell and a little background on people involved in the stories. The book is mostly based on discoveries made in north-west European shelf seas in the last 200 years. It so happens that many of the most important discoveries about how shelf seas work were made in these waters. That sets the geographical context but, in fact, the processes are not exclusive to here; they are relevant to shelf seas everywhere. Most of the science is physics, because that is my subject, but I have included some chemistry, biology and geology where I can. It is the nature of the science of the sea that each of these scientific disciplines is important to the complete story.

A book with the title *Home Waters* should explain whose home we are talking about. The seas on the continental shelf of north-west Europe are bounded, on the continental side, by Norway, Denmark, Germany, the Netherlands, Belgium and France. Sitting on the continental shelf and bathed in these waters are the islands – thousands of them – that belong to the Republic of Ireland and the nations of the United Kingdom (as well as the Channel Islands and the Isle of Man). What to call these islands? Not everyone, it has to be said, is happy about the name 'British Isles', but it is a convenient and ancient one; it dates back to a time before modern political divisions. Mostly in this book I have used this name for the sake of brevity, but sometimes you will see other names such as the British and Irish islands.

CHAPTER 1

A QUESTION OF DEPTH

W HEN IT WAS MY TURN TO interview prospective students of oceanography, I would sit them down, make them a cup of tea and ask how deep they thought the ocean might be. Not many knew the answer with any confidence, but that was alright. The purpose of the question was to get a conversation started and it usually worked. The short answer is that salt water fills the ocean basins – the Pacific, Indian and Atlantic – to an average depth of about 4km. There are mountains on the ocean floor, some of which rise to the surface, and canyons, some reaching down to two or three times the mean depth, but much of the sea floor – the abyssal plain – is quite flat and has a depth close to the mean figure. This brief answer, however, is not the end of the matter. As far as this book is concerned, it is only the beginning.

The ocean basins are presently overfilled. Water has spilled out to cover the low-lying edges of the continents, an area known as the *continental shelf.* Shallow seas, called *shelf seas,* lie around the rim of each continent. The average depth of the shelf seas is just a few tens of metres, a small fraction of the great depths of the adjacent oceans. In total, the shelf seas are small in area (and even smaller in volume) compared to the deep ocean, but their close proximity to land enhances their importance. For nearly everyone on the planet, the nearest body of salt water is a shelf sea. It is shelf sea water that splashes our feet as we walk along the edge of the beach and shelf seas that crash against cliffs during a winter gale. The deep ocean may be far grander, but it lies over the horizon and out of sight.

Shelf seas are more than shallow and small extensions of the ocean; they work in different ways and other processes are important. Fresh water flows into the sea from the land. The presence of the continental shelf amplifies the ocean tide; tides and tidal currents are greater and faster on the shelf than in the ocean. The shallow water, stirred by tides, is biologically productive; the most important fisheries on our planet are all found on the continental shelf. The rich biology makes the sea green, not blue, and the water sparkles with millions of flakes of clay mixed up from the seabed. Because they are shallow and close to land, shelf seas offer commercial opportunities for mining mineral resources on and below the sea floor. We are also learning, slowly, how to extract the energy contained in the tides and waves on the shelf and use it to provide electricity for our homes and industries.

As we look to reduce our dependence on fossil fuels, the energy contained in shelf seas will become increasingly more important. Shelf seas are full of movement and variety and they are accessible. They may be small compared to the ocean but in their economic and social value they punch above their weight.

The islands that compose Britain and Ireland sit, surrounded by shelf seas, on the continental shelf of north-west Europe. The area of the shelf, between the latitude of Brittany in the south and Bergen in the north, is 1.2 million sq km, about four times the land area of the Irish Republic and United Kingdom combined. It is larger than any European country west of Russia. The dimensions of this shelf make it particularly good at amplifying the tide – some of the largest tides in the world (and the fastest tidal currents) are found here. The water depth is mostly less than 200 metres; two-thirds, by area, has a depth less than 100 metres; one-third is shallower than 50 metres. It would be difficult to find a place on the shelf where you could submerge Blackpool Tower, which is 158 metres high; the Empire State Building (381 metres) would have plenty of room on top for a giant ape, wherever you put it.

Beginning at the shore, the continental shelf slopes gently down towards the ocean. There are features on the seabed – hills and valleys – that may need keeping an eye on as navigational hazards or places that offer good fishing potential, but these are generally small affairs compared to what we can see on land and the drama that lies ahead. At the edge of the continental shelf, the seabed starts to tilt more steeply downwards. The beginning

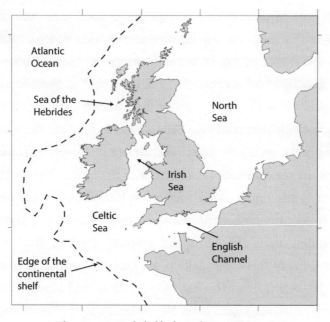

The continental shelf of north-west Europe.

of the steeper slope is called the *shelf break* and beyond this the sea floor descends at a gradient of about 1 in 10 to the ocean floor, the abyss, thousands of metres below. This gradient, the *continental slope*, is the true edge of the continents. It is one of the most prominent features on the surface of the solid Earth but it is largely unknown because it lies unseen, covered by ocean.

If we could stand on the floor of the ocean and look at the edge of the continental shelf, we would see a sight unmatched by anything above sea level. The sea floor rises before us, from a depth of several kilometres to within almost touching distance of the sea surface. It does so up a slope that stretches to left and right for thousands of kilometres. The shelf edge twists and turns as it

4

curves around the continents and there are deep canyons cut into it. Down these canyons, dense water tumbles from the shelf seas into the ocean carrying with it oxygen and carbon and the detritus of the land. As the tide flows up and down the continental slope it squeezes and then stretches the water column like an accordion, setting waves shimmering on the different density layers within the sea. Remarkably, there is little sign on the surface of the ocean of this large underwater structure. A ship could cross it without knowing, but the shelf edge is an important boundary around our islands, nonetheless. It affects the way that waves, tides and currents travel from the deep ocean to the shore.

Nautical charts of western Europe in the early part of the 19th century show an abundance of depth measurements on the continental shelf and none at all in the ocean. The shelf edge is marked as a dotted line with the tantalising caption, '*The soundings outside of this dotted line are very deep*'. The lack of observations in deep water is understandable. The technology of the time, a weight at the end of a marked line, worked well enough in shallow water but was extremely difficult to use with any accuracy in the ocean. Moreover, there was no practical need to know the depth beyond the continental shelf. Ships' captains were only interested in the water depth if they intended to anchor or were concerned that their vessel might run aground. No one knew, for certain, how far you had to go down the continental slope to reach the bottom. It was an interesting question but there was no practical incentive, at the time, to answer it.

That all changed with the proposal to lay an electric telegraph cable along the floor of the Atlantic Ocean between Europe and America. The first underwater telegraph cable had been laid across the English Channel in 1850 and there were bold plans to follow this achievement with a cable across the Atlantic. The success of the project relied on knowing not just how deep the ocean was but also about any obstacles on the planned path and the best way for the cable to negotiate the continental slope. In 1853, Lieutenant Commander Berryman of the US Navy took a series of depth measurements and bottom samples along a line from Newfoundland to Ireland, and in the summer of 1857, HMS *Cyclops* of the Royal Navy was dispatched to make deep-sea soundings along the proposed route of the cable.

I was curious to know more about the *Cyclops* survey and made an appointment to visit the archives section of the United Kingdom Hydrographic Office, where there is a copy of the cruise report. After passing the security checks at the gate, visitors are taken to a room in the archives building where they can spread out maps and books on a long table. There was only one other visitor when I was there, a historian from Oxford. The *Cyclops* report is contained in a slim pamphlet, *Deep Sea Soundings in the North Atlantic Ocean between Ireland and Newfoundland*, written by Lieutenant Commander Joseph Dayman. The *Cyclops* was a paddle steamer, with great wheels on either side driven by its 320 horse-power steam engines. In April 1857, Dayman commissioned his new ship 'for special purpose' at Sheerness in Kent and took it for trials of the sounding gear in the nearby River Medway. Dayman's report is clear and to the point and the writing style makes it a pleasure to read. 'My orders', he writes, with admirable understatement, 'were simply to carry a line of soundings from Valencia in Ireland to Trinity Bay, Newfoundland.'

The techniques used in the earlier survey by the Americans were adopted and modified for the *Cyclops*. Special deep-sea sounding lines were used with a heavy detachable weight, released by a mechanism that operated when the weight touched the sea floor. The sudden release of tension on the wire when the weight was dropped made it easier to be sure that the bottom had been reached; it also saved having to bring the heavy weight back to the surface.

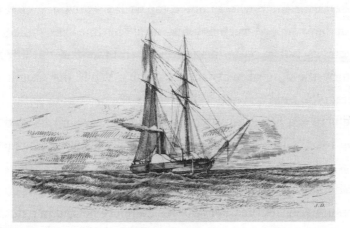

HMS Cyclops.

The *Cyclops* steamed out of Valencia in early June 1857, its great paddle wheels churning the water of Dingle Bay. The ship reached Trinity Bay in Newfoundland three weeks later and then steamed back to Ireland, filling in gaps in the survey line along the way. The ship's track in both directions lay along a great circle route: the shortest distance between two points on the surface of a globe; it was the same route as that planned for the telegraph cable.

The procedure at each sounding station was the same. The sounding line was paid out over a davit on the bow and the heavy weight was allowed to fall as quickly as it could pull out the line. In deep water it could take the best part of an hour for the weight to reach the bottom and the crew had to stay alert during this time, or they could easily miss the crucial moment when the weight hit the sea floor. The line was marked at 50-, 100- and 1,000-fathom intervals and the time that each of these passed over the bow was noted (apart from anything else, this must have helped keep interest up). The line was pulled out by gravity acting on both the weight and the paid-out line and also by the drag of subsurface currents. While the weight was falling, the intervals between fixed marks passing over the bow might be expected to reduce slowly as the weight of, and the drag on, the line gradually increased. When the weight hit the bottom and was no longer pulling the line down, there would be a sudden increase in the time interval between fixed marks. The measurements of the interval were therefore helpful in fixing the moment that the weight hit the bottom. I think I would also have tested the tension in the line to see if I could tell if the weight had been released, although there was no mention that they did this, so perhaps the change was undetectable.

The crew had to be sure that the line was properly vertical. Wind blowing on the ship would move it sideways so that it was no longer directly above the weight. This effect could be reduced by passing the line from the davit on the bow through a buoy floating on the surface. The wind effects on the buoy would be smaller than those on the ship and, in the absence of subsurface currents, the line would hang vertically from the buoy. The job of the ship's crew was then to use the engines and sails to keep the ship's bow as close as they could over the buoy. They soon became skilled at doing this so they could be pretty sure the line was vertical as it entered the water.

There remained the problem of bending of the line by subsurface currents. The extent to which that was happening could be tested by comparing the depth measured with the line and an independent measure of the depth made with an instrument called a Massey's sounding machine. This machine was placed directly on the sounding line, just above the detachable weight. As the sounding machine sank to the seabed, the turns of a rotating vane were counted on a set of dials, which could then be calibrated in terms of the fathoms descended. The results from the line and the sounding machine (corrected by a calibration carried out by Dayman) were close enough to conclude that the effect of subsurface currents was not a serious problem to the accuracy of the soundings.

The report includes remarks on conditions at each sounding station. For example, here are the notes from a station at 52°29'N, 26°14'W:

July 18. Homeward voyage. Fresh breeze at W. by S. [7.].
High Sea. Bar 29.820. Temp. of air 63°, of sea 58°.
Weight employed 96 lbs iron, detaching apparatus and
tapered whale line.
[This sounding was made in a fresh breeze with a high
sea and was attempted only for the purpose of filling in a
deficiency left going out through bad weather.]

There follows a list of times of the 100-fathom marks on the cable being passed out over the bow. It took between two and two and a half minutes to pay out 100 fathoms of line and this interval increases markedly when the weight hits the bottom. Then comes a comparison between the depths measured by the sounding line and Massey's machine (in this case, in 2,400 fathoms the agreement was within 50 fathoms) and a note about sediments collected from the sea floor: '*Oaze in plentiful quantity*'. These bottom samples were stored in bottles for later analysis.

It looked to Dayman that the greatest obstacle to laying the telegraph cable would be the steep slope at the edge of the continental shelf; as Dayman describes it: '*the steep declivity near the Irish coast, where in a little more than 10 miles of distance a change of depth occurs amounting to 7,200 feet.'* This gradient was the steepest encountered on the voyage (on a survey a few years later, Dayman reckoned that the continental slope further south, at the latitude of the English Channel, was somewhat steeper). I've used the *Cyclops* soundings to draw a profile of the edge of the shelf in the sketch below. As is usual with these figures, the vertical scale has been exaggerated so that it can be seen at all. The dip in the bed on the continental shelf (under the scale bar) is there because the *Cyclops* caught the edge of the deep water that protrudes on to the shelf to the south-west of Ireland (you can see this incursion of the shelf edge on the map on page 4).

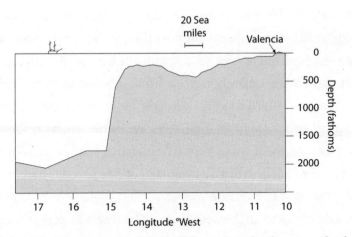

The profile of depth westwards from Valencia in Ireland out into the deep ocean, as measured by HMS Cyclops.

Cyclops' sounding gear was fitted with a device to bring back samples of the seabed. On the shelf, the material brought to the surface was sandy or 'fine sand' (a muddy sand). On the slope it was a mixture of sand, mud and rock. In the deep ocean beyond the slope, the bottom was, Dayman writes, '*a soft mealy substance, which, for want of a better name, I have called* oaze. *This substance is remarkably sticky, having been found to adhere to the sounding rod and line through its passage from the bottom to the surface.*' Today, we would probably call this material *ooze,* a slight change to Dayman's choice. The ooze samples were stored in bottles, which were returned to London for examination by the biologist Thomas Huxley. Huxley was, at the time, one of the most eminent biologists in the world. He was known as 'Darwin's bulldog' because of his vigorous and eloquent support for the theory of evolution. The samples of deep-sea sediment from the *Cyclops* must have been something entirely new to him – the equivalent, today, of samples of Martian soil – and yet he seems to have been a bit sniffy about them, at first anyway. In his initial response, appended to the *Cyclops* report, he writes that he hasn't had time to do much with the samples yet because he had '*many more pressing and immediate duties*'.

Huxley was to make something of these samples, though. Under the microscope, he found large numbers of 'very small rounded bodies', to which he gave the name coccoliths (roughly translated as 'grains of stone'). It was these that were giving the ooze the mealy quality that Dayman had noted. The coccoliths dissolved easily in dilute acid; they were made of calcium carbonate or chalk. Coccoliths are the outside skeletons of a type of microscopic algae that live in the surface of the ocean. When these particular algae die, their chalk coats sink to the ocean floor. We now regard this as an important part of the ocean carbon sink: the sea absorbs carbon dioxide from the atmosphere, algae remove carbon dioxide from the surface layers of the ocean during photosynthesis and the carbon sinks in small solid particles, such as coccoliths, to the ocean floor where, after a very long time, it becomes sedimentary rock. In the case of coccoliths, the rock that is produced is chalk – the very chalk, in fact, that makes the White Cliffs of Dover.

There was great interest among biologists of the time about whether life could exist in these great ocean depths. The closest *Cyclops* came to answering this question was when she brought up some shell fragments from a depth of 1,765 fathoms (3,228 metres). The question was settled a few years later when another deep-sea sounding survey, by HMS *Bulldog*, brought up several live starfish, their slender arms curled around the sounding line, from a depth of 1,258 fathoms (2,300 metres).

The voice of my host, John Williams, brought me back from these issues of mid-19th century biology to the present day. 'Time for lunch,' he said. John took me for a walk in the late autumn sunshine. There were brand new offices to be seen on the site, empty at the time of my visit as people were working at home during COVID. John showed me the original Hydrographic Office building in Taunton – the Edgell Building – used to house staff when they moved here from London in the 1940s to avoid the bombing. This building is shaped like a ship – John thought a liner, I thought it could be a battleship – with curved walls and floors layered like decks. This remarkable structure was also fitted with a funnel originally, although this had been removed when I saw it. Sadly, the building is no longer needed and, at the time of my visit, it was earmarked for demolition. Later, John showed me a drawing of the building in its heyday and I have reproduced it below.

Back in the archives room, I spread out a series of charts to see how knowledge of the shelf edge developed. 'Chart number 1' of western Europe appears in several versions. The first, an 1810 edition, shows relatively few soundings. There is some

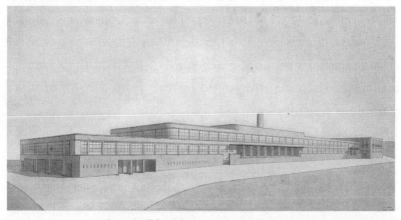

The Edgell building as it looked in 1944.

indication of where the water becomes deep to the west of the English Channel but the progress of the shelf break west of Ireland is not marked at all. The next edition of this chart 'with additions to 1830' shows more detail and the shelf edge is now marked as a dotted line, although there is still no knowledge of depths beyond the dotted line. The following edition, with corrections up to January 1866, shows deep soundings, including those of the *Cyclops* and the *Bulldog*, and the caption about the waters beyond the dotted line being 'very deep' has disappeared. The depth of the ocean beyond the shelf break was no longer a mystery.

Moving water can sometimes behave in unexpected ways. One day, my wife, Faith, and I were sitting in the car, parked in a high moorland lay-by, when snow began to fall. It was the sort of sleety snow that is miserable to be out in without offering any of the joys of crispy snowflakes. The drops that fell on the windscreen in front of us travelled down the glass in zigzags rather than straight lines. We both noticed this and remarked upon it. The drops waddled down the screen. We had been watching a programme on TV the night before about penguins and the motion of the drops reminded us of a penguin's gait. Each drop, as it moved downwards, became smaller as it left some of its mass in its watery wake. As the drop size reduced, the extent of the side-to-side movement also became less. I don't think this waddling motion happens with ordinary raindrops, which travel in straighter lines down the screen. It might be something to do with the nature of the water in the almost-frozen drop, but I don't know what that is.

The reason for mentioning this is that, on a much grander scale, the flow of water in the ocean can also behave in a non-intuitive way. At the beginning of the 20th century, a number of people were thinking about what might happen to the currents of the deep ocean – the Gulf Stream, for example – when they encountered a submarine obstacle such as the continental slope. The conclusion they came to was that the current would not be able to travel straight from the deep ocean on to the continental shelf (any more than the drops of sleety snow were able to travel straight down our windscreen). Instead, ocean currents that start to feel the ocean floor rising to meet them will turn until they are following the depth contours – flowing along the edge of the shelf instead of across it. Even the uppermost part of the current, which appears as though it would fit on to the shelf, chooses to turn with the rest of the water and follow the bottom contours.

There is a common view (still expressed in travel documents I saw when researching this book) that parts of the British Isles are bathed by the waters of the Gulf Stream and its extension, the North Atlantic Drift, but in fact very little of this great ocean current makes it on to the continental shelf. Instead, observations in the deep ocean show that the North Atlantic

Drift turns northwards when it reaches the edge of the shelf, travelling around Ireland and Scotland on its way to Norway (and beyond). Observations *on* the shelf tell us that there are no strong, steady currents flowing in any direction. Except for a few special places, the mean flow rate on the continental shelf (after tidal flows have been removed) would do a snail proud. For example, water entering the southern entrance of the Irish Sea emerges from the northern end about one year later. It travels a distance of 100km in one year at a mean speed of about one third of a kilometre a day or 3mm per second. The time taken for water to circulate around the North Sea is even greater – about four years. These times and speeds are not what we would expect if there were a great ocean current flowing across the shelf.

It is undeniably true that the British Isles are warmer than Labrador, which lies in the same range of latitudes on the western side of the Atlantic, but the contribution of the Gulf Stream and the North Atlantic Drift is mostly indirect. The warm current, even though it doesn't get on to the shelf, still travels close to our islands. The prevailing south-westerly winds blowing across the Atlantic towards us will be warmed by the ocean and bring its heat to us, indirectly. Another, local, factor is that the large shelf seas around our islands, stirred by the strong tides, are a great store of the summer heat of the sun. They release this heat in winter and prevent us from experiencing the prolonged freezing conditions seen in Labrador.

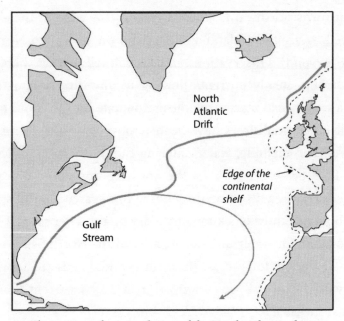

The main surface circulation of the North Atlantic skirts around the continental shelf.

The two people whose names are most closely associated today with the idea that major ocean currents that are deep enough to feel the bottom will tend to follow depth contours were contemporaries, Geoffrey Ingram Taylor (1886–1975) and Joseph Proudman (1888–1975). We will come across Taylor again in Chapter 6, but let me tell you a little about Proudman here. He was born into a farming family near Bury in Lancashire, attended Widnes Secondary School (where he showed an aptitude for mathematics) and spent nearly all of his working life in the north-west of England, becoming professor of mathematics and then professor of oceanography at the

University of Liverpool. Proudman turned out to have a flair not just for doing and teaching science but also for organising it. In 1919, he persuaded a local shipping company, the Booth line, to fund the establishment of the Liverpool Tidal Institute and became its first director. Later, along with Vice-Admiral Sir John Edgell (after whom the fancy building in Taunton was named), he was largely responsible for establishing the National Institute of Oceanography. Both these organisations exist today, although their names have changed. Much of Proudman's scientific work was concerned with the dynamics of the ocean – the way that water responds to the forces acting upon it. He gives a short and clear account (without too much maths) of his thinking about the way ocean currents might be steered by the shape of the seabed in his book *Dynamical Oceanography*, published in 1953 (although he had the original idea long before, when he was in his early twenties).

To understand why the currents of the Atlantic are deflected by the continental slope, we need to take a moment to appreciate how fluids behave on a spinning Earth. Moving objects (such as winds and ocean currents) on the Earth's surface appear to be deflected to the right in the northern hemisphere and to the left in the southern hemisphere. I say 'appear to be' because this deflection would not be seen by someone watching from a fixed position above the Earth. You can demonstrate this effect with a simple experiment. Cut out a circle of paper and attach it to a board with a pin through its centre. Now place a ruler across the paper and ask someone to turn the paper as you draw a line along the edge of the ruler. The pencil will trace out a curve on the paper, even though it is moving in a straight line along the ruler's edge. The fact that the pencil is moving over a turning surface makes its path appear curved to someone who is plotting out the motion on that surface.

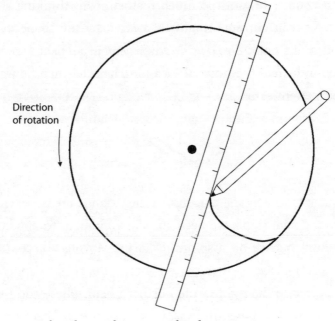

Direction
of rotation

A line drawn along a straight edge appears as a curve
on a rotating base.

This apparent deflection of moving objects on a spinning base is called the *Coriolis effect*. To account for its influence on ocean currents and winds we imagine a force, the Coriolis force, acting at right angles to the direction of the motion, towards the right in the northern hemisphere and to the left in the south. The strength of the force is proportional to the speed of the motion and it varies with latitude from zero at the equator to a maximum at the poles.

As the Gulf Stream and then the North Atlantic Drift cross the ocean, the Coriolis force pushes water to the right of the direction off flow. This sideways pushing continues until the pressure of the extra water on the right of the current matches the Coriolis force. The flow then comes into a balance, or steady state, pushed equally from both sides. If you've ever been in a large wide queue on your way to a sports stadium you will have felt something like this. People press against you from left and right but these forces cancel each other and the queue moves steadily forwards.

The balance between the Coriolis force and a difference in pressure is an important one in the atmosphere and the ocean. It is called a *geostrophic balance*. In the case of surface currents in the ocean, the pressure difference is produced by a slope on the sea surface, up towards the right of the current in the northern hemisphere. This slope produces a force within the water called the *pressure gradient force*. We will return to this force later in this book, but all we need to know now is that it acts horizontally in the water, in the downslope direction, and it has a strength proportional to the steepness of the slope on the sea surface.

We could draw contours of sea surface height and the current would flow between these contours without crossing them, just like winds blow between lines of constant pressure, or isobars, on a weather map. At each point in the flow, the Coriolis force acting to the right of the current is balanced by an equal pressure gradient force acting to the left. If the current speeds up and the Coriolis force increases, the slope on the sea surface responds by becoming steeper and the contours of sea surface height move closer together (in the same way that contours of land height on a map crowd together where the slope is steep). Conversely, if the current slows down the contours move further apart. So long as the latitude (and so the Coriolis effect) doesn't change too much, there is an inverse relationship between the speed of the flow and the spacing of the contours of sea surface height.

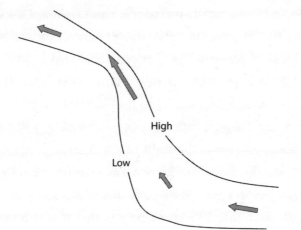

A geostrophic current flowing between contours of sea surface height speeds up where the contours are close together.

The current flowing between two contours of sea surface height carries water from one place to another. The volume of water that passes a fixed point each second is equal to the current speed multiplied by the cross-sectional area of the flow. We call this quantity the volume flow rate, or just the flow rate. In a current that is steady (that is not changing with time), the volume flow rate is the same in every part of the flow. This must be so, otherwise water will build up in one part of the flow (and become depleted in another) and the flow will not be steady. Steadiness is an important quality of ocean currents – the sort that you might want to draw on a map, at any rate. If the current is not steady, it will change and it will keep on changing until it does become steady. The steady state therefore represents a sustainable condition, one in which the current can flow along a more or less set path at a more or less set speed (the 'more or less' is needed because there will always be fluctuations produced by turbulent processes).

In his book *Dynamical Oceanography*, Proudman calls the conclusion that ocean currents deep enough to feel the seabed will follow depth contours rather than cross them 'a remarkable relationship'. His thinking goes like this. He imagines a current contained between two adjacent contours of sea surface height and reaching right down from the surface to the seabed. The cross-sectional area of the current is equal to the distance between the contours multiplied by the depth of water. If we call the separation of the contours s and the depth d, the cross-sectional area is s times d. The speed of the current in geostrophic balance will depend inversely on the spacing of the contours; it is proportional to $1/s$. When we calculate the volume flow rate by multiplying the cross-sectional area by the current speed, the contour spacing, s, cancels; the flow rate depends only on the depth, d. If the flow rate is to remain constant (as it must in steady state), the current is forced to stay in the same depth of water. It cannot move up or down a slope. Remarkable, indeed.

Geostrophic currents in steady state will either follow depth contours or stay in water deep enough so that they cannot feel the bottom. This explains why the deep-water currents of the North Atlantic do not reach our islands. When they arrive at the continental shelf, they are deflected to flow north or south, following the contours of the continental slope. There are many examples where currents at the surface are seen to follow the shape of the contours of the seabed: the flow is said to be topographically steered. The principle works in shelf seas too. On page 88 there is a map of the water circulation on the north-west European shelf. A current flowing down the east coasts of Scotland and England is deflected to the east when it reaches Yorkshire and it then flows across the North Sea to Denmark. I suspect it is following the 50-metre (165-foot) depth contour that lies across the North Sea at this latitude.

There are exceptions to the rule connecting currents and depth. Tidal flows, for example, which accelerate back and forth and are not steady, can cross the shelf break. Bottom flows, which

24

are affected by friction at the seabed (and so are not geostrophic), can also cross the boundary. One of the important ways that material travels from the continental shelf to the deep ocean is in these bottom flows flowing like avalanches down the continental slope.

There is a special feature of geostrophic currents flowing between lines of constant sea surface height. If the water is homogenous (that is, of uniform density), the pressure gradient force is the same from the surface down to the seabed (or to a depth where the water is no longer homogenous). Since the pressure force is balancing the Coriolis force, which depends on the current, the current must be the same all the way down to the seabed. If the current changes direction (or speed) at one depth, then it will change at all depths. It is as though the flowing water is locked together and has to follow what the water above and below is doing.

Sometimes, a current comes across an isolated mountain on the floor of the ocean. When it does so, it will tend to curve around the sides of the mountain, following the bottom contours, rather than flowing over the top. In a homogenous ocean in geostrophic balance, the contours of sea surface height and the flow at all depths bend around the submerged mountain. It is as though a phantom mountain sits on top of the real one, deflecting the flow right up to the surface. The column of water above the submarine mountain, around which the flow bends, is called a Taylor column, after GI Taylor.

You can see videos on the internet of laboratory experiments demonstrating the formation of a Taylor column above a submerged object in a rotating flow. My favourite is by the excellent Dr Mirjam Glessmer, who describes how to do the experiment with simple equipment and is honest about her failures. Actually, I don't find the laboratory demonstrations very convincing, but I am told that on the large scale in the ocean there can be a real effect (although it may not reach right to the surface because the ocean is not homogenous enough). It has been suggested that the Great Red Spot in Jupiter's atmosphere is an example of a Taylor column above a mountain on the planet's surface that we cannot see. There are other possible explanations of Jupiter's Red Spot, but the Taylor column idea is certainly a possibility. I find it quite moving to think that an idea developed by two young people (Taylor and Proudman) over a century ago can be used to explain an atmospheric feature on a planet no one has ever been near. It is a testament to what we can achieve with our brains when we try.

There is one final thing to add about the deflecting effect of the Earth's spin. It only becomes noticeable if the movement of the water lasts long enough for the turning of the Earth to matter. Ocean currents, which flow steadily over long periods are certainly influenced by it, but the spinning motion of water down a plughole, which lasts for just a few seconds, is not. Sadly, the idea that water spins its way out of the sink in different directions either side of the Equator is not true. Instead, the way the water turns when you pull the plug depends on how you have used the taps to fill the bowl. The taps are usually off-centre and if the flow out of one tap is a bit stronger than the other it creates a circulation in the bowl. When you pull the plug, this circulation becomes concentrated in the lovely spiralling motion in the water disappearing down the plughole.

You might be thinking that a major structure like the edge of the continental shelf must have an effect on the biology of the ocean, and you would be right. Some animals have found a niche in the canyons cut into the continental slope, where they benefit from dense water carrying material downhill to the deep ocean. The gradient in depth at the shelf edge also creates a unique ecosystem, which was first detected from space. Satellites have told us a lot about the way shelf seas work as we shall see in later chapters, but one story belongs here because it concerns the shelf break. In the 1970s, high-quality satellite infrared pictures first revealed the presence of a band of cool surface water following, exactly, the edge of the continental shelf from northern France around the coast of Ireland and up to the north coast of Scotland, a distance of 1,500km (932 miles). This was not something that anyone had anticipated, and it demanded an explanation. Satellite images of ocean colour showed that the band of cool water is also greener than the surrounding sea; the green colour is made by large numbers of the tiny photosynthetic organisms, phytoplankton, which are clearly finding the right conditions to grow along the shelf break.

The band of biologically productive water along the edge of the shelf is caused by the disruption of the natural summer layering of the sea as it makes the transition from the deep ocean to the continental shelf. In summer, the sun's heat warms a surface layer that sits on top of deeper, cooler water. The sea is said to be *stratified* and the interface between the layers of warm surface water and cooler deeper water is called the thermocline. Tidal motions create waves, called internal waves, on the thermocline. These waves break on the edge of the continental shelf in the same way that surface waves on the sea break on the sudden shoaling of the seabed at a sand bar. As the internal waves break, water from below the thermocline is mixed to the sea surface, creating the cool water that we see in the satellite pictures. The breaking internal waves also transfer nutrients from beneath the thermocline to the surface. These nutrients are the fertilisers of the ocean: they include nitrogen, phosphorus and other elements essential to the growth of marine phytoplankton. In the early part of the summer, nutrients are used up in the surface layer by growing phytoplankton; photosynthesis in a stratified ocean is then limited by the rate at which nutrients can be supplied from deep water to the surface. In most places, this upward transfer is slow: the thermocline acts as an effective barrier to vertical mixing. But at the shelf break, the breaking internal waves make good mixers. Phytoplankton growing in the surface layer at the edge of the continental shelf are fertilised from below and receive sunlight from above. They thrive and produce the green band, which is clear enough to be seen from space.

The plentiful food supply attracts spawning fish to the shelf break: they lay their eggs here so that the young fish can feed on the phytoplankton. The small fish attract larger predators. Basking sharks, which eat fish larvae, have been tracked to the shelf break where they are seen to home in on the concentrated patches of food. This is a fine example of an ecosystem created by particular physical conditions: a steep step on the floor of a tidal, stratified ocean.

The scientific study of the sea – now usually called oceanography – is generally acknowledged to have begun with the voyage of a specially commissioned research ship, HMS *Challenger*, in the years 1872 to 1876. The *Challenger* made the first global survey of the biological, chemical and physical properties of the world's oceans. The report of the Challenger Expedition can be found on the internet and it is impressive in its thoroughness. It runs to 50 volumes, which cover just about everything to do with the cruise (if you want to know where the toilets were located on the *Challenger*, there is a diagram to show you). The first two volumes are a narrative account of the voyage and are particularly fascinating. They are a mixture of travelogue, scientific information and navigational details. There are well-told stories about the places they visit and the people they meet. These volumes capture the adventurous spirit of an expedition that was as much a journey into the unknown as a flight to Mars.

The report contains an engaging account of the discovery of the deepest part of the ocean, now known as the Challenger Deep. The ship had arrived at the location by chance and the amount of sounding line needed to find the bottom took the scientists and crew by surprise. Two thermometers attached to the line were broken by the great pressure and, the report tells us *'Mixed with mud brought up by the* (bottom sampler) *on this occasion was some mercury out of these broken thermometers, which, falling faster than the sinkers, reached the bottom first, and, owing to the perfect stillness of the sea at this great depth, was caught by the rod descending exactly on the spot where the quicksilver had fallen.'*

These exciting discoveries in the great depths set a precedent. The motivation for subsequent voyages of scientific discovery was to improve the understanding of the *deep* ocean. The shallower waters around the coast – the shelf seas – were considered, in comparison to the great oceans, to be of lesser importance and not so interesting. It took a few pioneers, working mostly in the early-to-mid 20th century, to change minds about that. They showed that the seas on the continental shelf are interesting in their own right. The scientific problems that they present are different but no less challenging than those of the deep waters of our planet. There was some catching up to do to bring knowledge of the shelf seas up to the same level as that of the oceans. This work would take several generations of observations and theoretical ideas. Much of the work, as it happened, would be done in the seas around the Irish and British islands. There were discoveries to be made. And there still are.

CHAPTER 2

COASTAL CURRENTS

THERE IS AN OLD FABLE TELLING how the sea came to be salty. Someone had invented a machine that manufactured salt and ran forever. The machine was sent as a gift and by sea to a neighbouring kingdom, but on the voyage the ship carrying the precious machine was caught in a great storm. The ship and its cargo sank to the bottom, where the machine continued to churn out salt, as it does faithfully to this day. I think a tale like this one features in the folklore of several countries and it is easy to see why a story would be needed. The saltiness of the sea is a puzzling thing. Where does the salt come from? Why aren't lakes – even very big lakes – salty like the ocean?

It is now believed that the salt in the ocean comes mostly from rivers flowing into it (with a lesser but more dramatic input from submarine volcanoes). The rivers pick up the salts as they flow over rocks and soil; river water doesn't seem particularly salty but in fact it contains trace amounts of most of the elements. Lots of rivers flow into the ocean and some of the water they bring evaporates from the sea surface, falls as rain on the land and ends up back in the rivers. Evaporation only removes fresh water, so the salts brought into the ocean by the rivers get left behind. Lakes don't normally become salty because, although rivers do flow into them too, rivers also flow out of most lakes, carrying the salt with them. The balance of water in a lake is different to that of the ocean. Water is added to a lake by rivers and removed mostly by other rivers. Lakes that are an exception to this rule, such as the Great Salt Lake in the United States, do become salty.

Within the ocean, the salts brought in by rivers and underwater volcanic eruptions are used in biological and chemical processes. The elements within the salts are removed from the ocean by sinking to the ocean floor to create sediments and, ultimately, sedimentary rocks. It is likely that the processes of adding salt to the ocean and removing it proceed at equal rates so that the salt content of the ocean is not changing. A simple analogy may help to see how this works. Imagine you are filling a bucket that has a small hole in it. Water squirts out of the hole at a rate that depends on the pressure 'head' – that is, the water level in the bucket. The higher the water level gets, the faster you lose water through the hole. Things will come into equilibrium when the level in the bucket is just right to make the water squirt out at the same rate as you are adding it. If the hole is a small one, the equilibrium level of the water in the bucket will be higher than that for a large hole. In the case of elements in the ocean, the level (or concentration) of each element will be set such that chemical and biological processes remove it at the same rate as it is added. Elements that are removed slowly will be present in higher concentration than the more chemically and biologically active elements. Sodium and chlorine (the components of common salt) are the most abundant elements dissolved in sea water not because they are present in the highest concentration in river water (they are not) but because these particular elements are removed slowly from the ocean.

The average salt content (or *salinity*) of the oceans is about 35 parts per thousand; that is, 1kg (2.2lb) of sea water contains about 35g (1¼oz) of salt. The salt content of sea water changes a number of its physical properties compared to fresh water. One is that sea water is *denser* than fresh water at the same temperature. Density is the mass of an object divided by its volume. The density of fresh water at 15 degrees C is 1,000 kg.m^{-3}; the density of sea water at the same temperature usually lies between 1,020 and 1,030 kg.m^{-3}. We sometimes express the higher density of sea water using the concept of specific gravity, the ratio of actual density to fresh-water density. The specific gravity of sea water is therefore normally in the range 1.02 to 1.03. One consequence of the higher specific gravity of sea water, which you might have noticed, is that it is easier to stay afloat in the sea than in a fresh-water swimming pool. We get more buoyancy from the higher density sea water.

The influence of salinity on the specific gravity of water sometimes has to be taken into account when designing objects to float on, or under, the water surface. The tunnel that now carries the A55 – the main North Wales expressway – under the River Conwy was constructed as a series of rectangular concrete tubes, which were laid end to end in a channel cut through the sand and mud of the riverbed. Each concrete tube was the size (and probably the weight) of a small oil tanker and it was temporarily sealed at the ends so that it could be floated into place. The tubes were designed to be, on their own, slightly denser than water. At the time of deployment, buoys were attached to keep the tubes afloat as tugs towed them into position. When they were in the right place, the buoys were removed and the tubes sank gently to the bed. This delicate operation required the density of the tubes to be just right, slightly greater than that of the water they were floating in. If the tube density was too low, the tubes wouldn't sink (which would have been very embarrassing, to say the least – these weren't the sort of thing you could poke down with a stick). If the tube density was too high, they might sink too fast and be damaged when they hit the bottom. The density of each tube could be measured exactly but the density of the water was changing all the time as the tide pushed salt water in and out of the estuary. The engineers made careful measurements of the specific gravity of the water at various stages of the tide so that they knew the right time to release the buoys. When the day came, this tricky operation fortunately went smoothly.

The concrete tubes of the Conwy tunnel were constructed by the side of the river in a great basin that had been dug out in the 1940s to make the floating Mulberry harbours used to transport troops and goods to France after the Normandy landings. Today there is a pub – The Mulberry – nearby, from where you can look over the estuary and contemplate the specific gravity of your pint.

Estuaries everywhere have one thing in common – they have the sea at one end and a river at the other; together these make a gradient of salinity and density along the length of the estuary. The difference in specific gravity between the ends tends to produce a circulation – called the estuarine circulation – in which salt water flows inland along the bottom of the estuary and lighter, fresh water flows towards the sea at the surface. This circulation creates layering (or stratification), such that at any point in the estuary the salinity increases from the surface to the bottom (although we can imagine that mixing will make the surface waters no longer completely fresh but partly salty and the bottom waters will not be as salty as the sea at the estuary mouth).

There is a nice laboratory demonstration of the estuarine circulation called the lock-gate experiment. A rectangular Perspex tank is filled with water and then divided into two sections by a vertical 'lock gate': a sheet of Perspex that slides down through grooves on the inside of the tank. Some kitchen salt is added to one side of the lock gate and stirred until it dissolves. A dye – milk, for example – can be mixed in with the salt to make the flows visible. When the lock gate is removed, the salt water flows, rather slowly and majestically, along the bottom of the tank and fresh water flows in the opposite direction over the top. There are some interesting effects when these currents reach the end of the tank but eventually the motion stops with a layer of fresh water lying on top of a salty layer. There are traces of milk in the upper layer, indicating that some exchange, or mixing, has taken place.

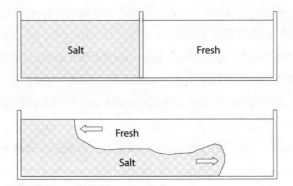

The lock-gate experiment.

The lock-gate experiment is an example of the conversion of potential energy to kinetic energy, the hydraulic equivalent of rolling a ball off the top of a hill. At the start of the experiment the centre of gravity of the salt and fresh water is at the same height, halfway between the surface and bottom of the water in the tank. At the end of the experiment, when the fresh water sits on top of the salt water, the centre of gravity of the fresh water has been raised but that of the salt water has been lowered by the same amount. Because the salt water is heavier than the fresh water, this means that there has been a net reduction in the potential energy of the tank.

The potential energy has been converted into the kinetic energy of the flow that is created when the lock gate is removed (ultimately, when the flow has stopped the energy is turned into heat through frictional forces on the flow and also a little bit of mixing between the layers, which restores some of the potential energy). We can work out the speed of the flow by matching its kinetic energy to the loss of potential energy. The speed depends on the depth and the density difference between the salty and fresh water. For a laboratory tank 20cm (8in) tall with fresh water at one end and typical sea water at the other, the flow speed is about 8cm per second. Applying the same rule to an estuary 4 metres (13 feet) deep, we would expect the water to flow at a speed of about half a metre per second, or 1 knot. That should be easily measurable.

In fact, it is not all that easy to observe the circulation in a full-sized estuary. This is because the tide pushes water into and out of the estuary twice a day at speeds that can be ten times greater than the estuarine circulation. To measure the gentler flows driven by the differences in density, it is necessary to first remove the effects of the tide by averaging them out. Measurements need to be made over a tidal cycle, the time interval of 12½ hours between one high water and the next (or between some other point in the tide and the next equivalent point). This is best accomplished from a small boat anchored in the deepest part of the estuary, with current meters and salinity probes lowered by hand over the side. A profile of readings is taken from the surface to the bottom at regular intervals. Tidal-cycle observations make for a long day but provide an opportunity to reflect on why you ever decided to become an oceanographer and why you didn't go to the toilet before you got on the boat. It's a great help on these occasions to have good company and a generous packed lunch.

Tidal currents in an estuary mostly flow either in (when they are called the flood) or out (the ebb). We can use an arithmetic sign convention for recording this: flood currents can be noted as positive, for example, and ebb currents as negative. When the currents measured over a full tidal cycle are added up, the positive flood currents will tend to cancel the negative ebb currents, so that the tidal flows average out close to zero. A current such as the estuarine circulation, however, which flows steadily in one direction (at a set depth), will not average out when this adding-up is done. Summing currents in this way over a tidal cycle therefore reveals the presence of the weaker, but steady, non-tidal or residual currents.

The measured speed of the estuarine circulation is not as great as we would expect it to be from scaling up the lock-gate experiment. The flows are slowed down by a process called tidal mixing. The fast tidal flows in the estuary create turbulence: parcels of water are exchanged between the surface and bottom waters in the estuary, slowing the flow in both directions. A parcel of water from near the bed, with momentum carrying it landwards, is transferred to the surface where it slows down the seaward flow. At the same time, a parcel of water from the surface is carried to the bottom and the transfer of momentum slows the flow there. The turbulence also mixes salt between the layers, increasing the salinity of the surface layer as it flows seawards and reducing that of the bottom layer as it moves inland.

In any estuary, or indeed anywhere where there is a variation of salinity from place to place, there is a contest between the effect of gravity tending to reduce the potential energy (as it does in the lock-gate experiment) and the tides mixing salt upwards and restoring the potential energy. The physical nature of the estuary – its circulation and stratification – will depend on which process is winning this contest. The balance of power will change with time; tides vary with a fortnightly cycle and stratification is more likely at a neap tide when the currents are weakest. There are circumstances also when the tide can work to create stratification. Tidal flows are slowed down near the seabed by friction. When the tide is flooding, it brings salty water into the estuary fastest at the surface. If the water gets saltier at the surface than it is at the bottom, it will sink down, creating convection currents, which add to the turbulence. On the ebb tide, however, the faster flows at the surface tend to put fresh water from inland on top of saltier water. Regular pulses of stratification are produced on the ebb, and these will normally (but not always) be removed on the following flood. This process, called *tidal straining*, was first observed in Liverpool Bay by Professor John Simpson, who made many important contributions to the study of shelf seas.

Salinity is measured routinely at sea by scientists working on oceanographic research ships. There is a fleet of ten or so specially commissioned research ships operating in Britain and Ireland, with more based in the European countries with coastlines along the English Channel and North Sea. Some of these vessels are suitable for long ocean voyages and polar environments but others are more comfortable operating on the continental shelf. Traditionally, these ships have been given names appropriate to their purpose of scientific discovery: *Endeavour*, *Celtic Explorer* and *Challenger*, for example. Thankfully, we were spared the name *Boaty McBoatface*, which was the popular choice when the Natural Environment Research Council conducted an internet poll for the name of its new polar research vessel. That ship was named the *Sir David Attenborough*, although *Boaty* was retained as the name for one of its robotic underwater vehicles.

When I told people who didn't know me so well that I would be off for a few weeks on a research cruise, the usual reaction was curiosity and a little envy. The word 'cruise' catches the imagination; we are used to seeing television adverts for ocean cruise liners and exotic river cruises. Scientific research cruises, in contrast, are hard work; ship time is expensive and the scientists on board feel the pressure to make every minute count. They, like the ship's crew, work watches around the clock. The scientific complement on a research vessel – at least, on the sort of cruises I went on – is a mixed one. There are biologists, chemists, physicists and geologists, each with their own programme of work to carry out. Some will be young research students on their first trip to sea and others old hands who have seen it all before. But everyone on board knows that they have been given a rare, possibly unique, opportunity and they will be trying their hardest to collect the best possible data.

The ship visits a set of 'stations' – selected spots at sea where scientific observations are to be made. Generally, the physicists have it easiest here. Most of the measurements they want to make can be done with instruments lowered into the sea,

43

which collect a vast amount of information in a very short time. Members of other scientific disciplines collect samples of sea water or the seabed, which they will work through on the voyage between stations, or store for later analysis back in the laboratory. Things happen at sea that you just don't see on land. On one voyage we were steaming out towards the shelf break with nothing, or so it seemed, for miles around us. Suddenly, a whale surfaced close by the ship, made a great splash with its tail and then dived away from sight. The boatswain – standing with me on the bridge and a man of steel if ever there was one – ducked at the shock of it and then smiled wryly at me as he realised the futility of his reaction. At least his reactions were working. I stood open-mouthed and rooted to the spot during the whole encounter.

Research ships are fitted with an instrument, called a CTD, to make routine measurements of electrical conductivity (C), temperature (T) and depth (D). Salinity is calculated from the conductivity, which depends on the salt content of the water. The CTD is lowered over the side of the stationary vessel and a profile, from the surface to the bottom, of temperature and salinity is made. A modern CTD will be equipped with additional instruments to tell us about chemical and biological properties of sea water.

CTDs became regular fixtures on research vessels in the 1960s, but before that there were other ways to measure salinity. Samples of sea water were collected and later analysed in a precision electrical conductivity meter in the laboratory (in fact this procedure is still carried out today to calibrate CTDs against a sample of standard sea water). Collecting a water sample from the surface is straightforward – a bucket tied to a rope could be put over the side of the ship – but collecting samples from below the surface required special equipment. Subsurface samples were collected in a special bottle, which could be closed (or 'fired') at a chosen depth. Several types were available. One, named after the National Institute of Oceanography in the UK, was a plastic tube, about 46cm (18 inches) long and 8cm (3 inches) in diameter, fitted with end caps, which were closed by a strong spring. The bottle was clipped to a steel wire (called the hydrographic wire), which was held over the side of the ship and kept near vertical by a great lump of lead at the bottom. You 'cocked the bottle' by opening the caps against their strong springs (keeping fingers out of the way) and clipping them into position. The bottle was then lowered to the desired depth.

To close the caps and collect a water sample, a 'messenger' was sent down the wire. The messenger was a brass cylinder with a groove cut into the side. The groove fitted snugly over the hydrographic wire and part of the messenger could be twisted so that it couldn't fall off; once set, you would send it down the wire with a flick of the hand. It could take a while for the messenger

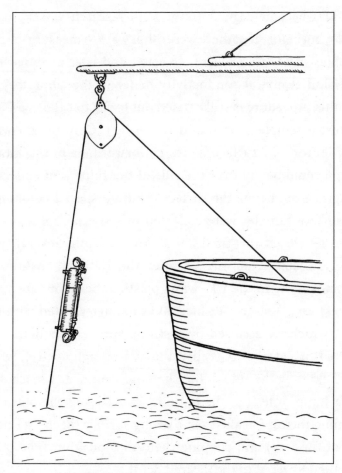

Collecting a water sample from depth.

to reach the bottle at depth and during this time you held the wire between fingers and thumb. When the messenger reached the bottle, it released the catch and the springs closed the caps; the vibrations caused by this action would travel up the wire to your fingers. It was quite something to stand there with the wind on your face, watching the green waves surging around the ship, waiting for the signal from the messenger.

Water samples from depth are still needed today for biological and chemical analysis, but now there is a ring of bottles arranged around the CTD and they are fired from a cosy and warm ship's laboratory by pressing a button and sending an electrical signal down the wire. I doubt many people mourn the passing of the old ways, but after nearly 50 years I can still remember the feel of the wire in my fingers as the messenger struck home and the quiet satisfaction of knowing that the job had been done correctly.

The average rainfall in the British Isles is about 1 metre (39 inches) per year, give or take some depending on where you are. Over a land area of 315,000 sq km, fresh water falls on these islands at a rate of about 10,000 cubic metres per second and flows towards the sea in rivers. When the river water gets to an estuary, it mixes with salt water, but it arrives at the coast still measurably less salty than the open sea. Because it is less salty it is also less dense than sea water and – if mixing by wind and tide is not too vigorous – it will spread out over the surface of the sea in what is called a plume. As it continues to spread, the effect of the Earth's rotation turns the plume to the right (in the northern hemisphere) and it flows along the coast in a *coastal current*.

Each of our islands has the potential to form a coastal current: a fringe of less salty water circulating around the island with the land on its right-hand side. Fresh water is added to the current from the rain landing on the island and, at the outer edge of the current, it mixes with the surrounding sea water. The width of the current – the distance it extends offshore – will then depend on the relative importance of these two processes. If there is a chain of islands close enough together for their fringes of fresher water to overlap, the current will pass from one island to the next. In reality, though, discernible coastal currents are only seen on coasts where there are large rivers (or a lot of small rivers) and where mixing at the edge is weak so that the current can maintain its coherence.

One place where a coastal current occurs is along the west coast of Scotland. The *Scottish coastal current* is formed from water leaving the northern channel of the Irish Sea, joined by more water travelling along the north coast of Ireland. As it flows north up the Scottish coast, this water is augmented by fresh water flowing in from numerous sea lochs. The Scottish coastal current is about 50km (30 miles) wide and has a salinity less than 34.5 parts per thousand. The salinity is only a little lower than that of the offshore water, but the difference is measurable. The current is able to jump over gaps between most of the Scottish islands but when it reaches a particularly large gap, between Tiree and Barra, it takes a detour to the east to make contact with the land again before it continues on its way. This low-salinity flow can be traced all the way up the west and north coasts of Scotland

and into the North Sea. Using the reduced salinity to track the flow is an example of using a *tracer*: something that identifies where the flow is. Flows that are very weak compared to the tide are difficult to measure. It is necessary to average over a large number of tidal cycles to find the residual circulation. That is why the idea of using a tracer in these circumstances is a good one; the averaging is done for you by the tracer, which has been in the sea for many tidal cycles. There are limits to what you can learn from a tracer, though: it might be able tell you where the flow is going but not usually how quickly it is moving.

The speed of the flow can be measured directly by recording current meters placed on moorings in the flow. This is the automated equivalent of the measurements of the residual flow in an estuary. For many years, recording current meters were electro-mechanical: they had an impellor that turned in the flow and the speed and direction of the current were recorded at regular intervals on a small magnetic tape. Today, current meters are more likely to be acoustic, using the Doppler shift in the echo of a ping of sound sent into the water to measure how quickly the water is moving. Recordings of currents (made over periods of months) in the Scottish coastal current show that the flow is, on average, northwards but it is unsteady with regular reversals in the direction of flow. The unsteadiness in the flow seems to be correlated with the passage of low-pressure systems to the north of Scotland. A strong north wind can stop or reverse the flow. This is a toddler among ocean currents, taking one step forwards and half a step back before sitting down.

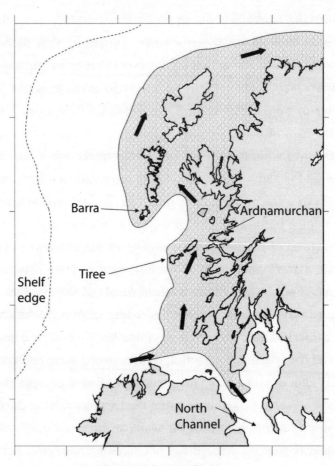

The Scottish coastal current.

The long-term average northward speed of the current is not great, a few centimetres per second or less than 10km per day. Nevertheless, this is one of the few places around these islands where there is a more or less persistent current in the same direction for most of the year. Multiplying the flow speed by the cross-sectional area of the current gives the volume flow rate – the volume of water

50

that passes a fixed point each second. For the Scottish coastal current, this figure is about 100,000 m³/s. On its passage northwards, the flow rate increases as fresh water runs off the land on the current's eastern flank and sea water mixes in from the west. By the time the current has reached the north-west corner of Scotland, the volume of water transported in the current is about twice as great as it was when it began its journey along the Scottish coast.

The west coast of Scotland is a particularly fine place for a cruise. There are interesting boltholes to escape to if the weather turns bad – good shelter behind an island or even better shelter in a sea loch. From the sea, the Hebrides appear as a procession of vivid green islands with inviting white beaches and turquoise water in the bays. On a good day, thousands of seabirds wheel and cry against a blue sky. One particular piece of land, dark and looming and fringed with white where the waves crashed against tall cliffs, caught my eye. 'That's Ardnamurchan,' a Scottish colleague told me, 'the most westerly point on mainland Britain.' Until then, I suppose I'd thought vaguely that Land's End was the most westerly point on the mainland, but the tilt of the country makes this part of Scotland reach further west than I'd realised. Most land viewed from the sea invites exploration and I resolved that one day I would see the Ardnamurchan peninsula close up.

It took me a long while to make this particular trip, but it was worth the wait. The journey by road from Glasgow takes you north along Loch Lomond, across the 'Rest and Be Thankful' pass and through Oban on the west coast. A small ferry between Corran and Ardgour sees you across Loch Linnhe and then there is a drive along the steadily narrowing A861, the only 'A' road I know that is 'single track with passing places'. The route passes through the small town of Strontian, where, in the 18th century, the unusual properties of the local ores led to the discovery of strontium (the only element named after a place in the United Kingdom). JM Barrie stayed not far from here when he was writing *Peter Pan*.

I pitched my tent at the village of Kilchoan, in a spot with views of the Isle of Mull across the water. A village post office and shop sells most things, including petrol. The filling station at Kilchoan post office is of a sort not often seen these days: it comprised a metal tank with oval ends sitting on a raised platform above ground; a perfectly sensible way of dispensing motor fuel in a small rural location. Petrol stations are few and far between in this part of the world and motorists have to be careful not to be stranded with an empty tank. On the day I was due to leave, I wasn't sure if I had enough petrol left, but the post office was also short and not due a delivery until later that day. They kept a small supply for emergencies, though. 'We can let you have ten pounds' worth,' said the young woman in charge. 'That should get you to Strontian and you will be able to fill up there.'

Before I left, I drove out to Ardnamurchan Point to stand as far west as is possible on mainland Britain. I knew that the foaming sea water in front of me was flowing, very gradually and unsteadily, northwards. Sometime before – perhaps 18 months ago – a fraction of this water would have been passing

Ardnamurchan lighthouse.

my home in Wales. I also knew that this water was slightly less salty than the water 80km (50 miles) offshore, although it was impossible to tell either of these things just by looking.

It is sometimes possible to estimate the speed of a coastal current from observations of salinity alone and by applying a rule known as the continuity principle. The idea was developed by the Danish physicist Martin Knudsen (1871–1949). Knudsen is unusual because he made his name in two quite separate branches of science: he is known for his work on the kinetic theory of gases at low pressure

53

but he also took part in the Danish *Ingolf* expedition to the North Atlantic in 1895–6 and became interested in hydrodynamics. His idea about calculating flow speeds from salinity measurements is so straightforward and powerful that I would like to show you how it works with some numbers, using the Irish Sea as an example. The Irish Sea, which lies between Ireland and the main island of Great Britain, is connected to the rest of the world's oceans by two channels, one in the north (called the North Channel) and one in the south (St George's Channel). Measurements made with a CTD in both these channels tell us that the water is vertically mixed. The temperature and salinity are the same from top to bottom (this happens because the tides here are very fast and tide-made turbulence stirs vigorously). The average salinity in the southern channel is slightly greater than that in the northern channel; the salinity is about 34.5 in the south and 33.5 in the north (see picture on page 55; salinity is usually expressed without units. We take it as parts per thousand or grams of salt per kilogram of water). Rivers bring fresh water into the Irish Sea; the combined input of fresh water from all rivers (mostly the Mersey, Ribble, and Lune) is about 2,000 cubic metres per second. Although we said earlier that rivers carry small amounts of salt into the sea, the concentration of these salts is negligible compared to that in salt water and we can consider the salinity of the river water to be zero. The river water has to escape from the Irish Sea somehow, either through the northern channel or the southern one, or a mixture of the two. A moment's thought tells us that the fresh water must leave through the North Channel. The water that leaves will be fresher than the water that enters the

Irish Sea, because of the river water that is added to it. So, the flow through the Irish Sea must be northwards, entering from the south and leaving to the north, picking up fresh water on its way.

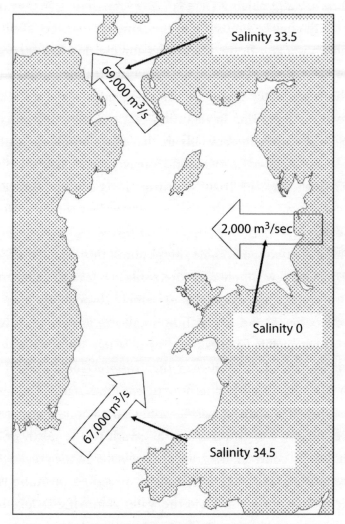

Water and salt budget in the Irish Sea.

The salt brought in to the Irish Sea through its southern entrance must match the salt carried out by the flow out of the North Channel. This means that the volume flow rates (the volume of water passing a fixed point every second) in the north and south channels must be in the same ratio as the salinities, namely 34.5/33.5 or 1.03, in order for the flows to carry the same mass of salt. The *difference* between the flow rates is equal to the fresh water input rate of 2,000 m³/s. We are looking for two numbers that have a ratio of 1.03 and a difference of 2,000 and a little algebra tells us that these numbers must be 69,000 and 67,000. Every second, on average, 67,000 m³ of salt water flows into the Irish Sea through its southern entrance and 69,000 m³ of slightly less salty water leaves by the North Channel.

There are two remarkable things about this calculation. The first is that a few simple measurements can lead to an estimate of flow rates that would be extremely difficult to measure directly. The second is just how small the answer is. Several tens of thousands of cubic metres of water sounds like a lot but if we spread this out over the cross-sectional area of the Irish Sea, the speed of the flow is less than 1cm/s The water is moving at a snail's pace; it would take about a year for a parcel of water to travel from the south to the north of the Irish Sea. By contrast, the transport of water in the great ocean currents such as the Gulf Stream is measured in millions of cubic metres per second and the water is moving at speeds of several metres per second. The residual flows out of the North

Channel of the Irish Sea are certainly nothing like the Gulf Stream, despite claims that the Gulf Stream touches this south-west corner of Scotland.

There are a number of places where a coastal current has been observed, at least for some parts of the year. One is on the coast of the Netherlands and Germany, where the River Rhine flows into the North Sea. Another is on the west coast of Ireland, where a coastal current has been tracked flowing northwards in summer (although this current may be driven by differences in temperature as much as salinity). The current turns right around the north-west corner of Ireland, the Bloody Foreland (so called because of the red colour of the rocks seen here at sunset) and then flows along the north coast of Ireland where part of it joins the Scottish coastal current but another part continues along the Irish coast and turns southwards into the Irish Sea. It hugs the coast of Northern Ireland, moving against the residual flow northwards in this channel. It is intriguing to think that the flow may continue around the rest of the Irish coast, making a complete loop. This is not impossible, but it would be difficult for such a current to keep going through the strong tidal mixing of the Irish Sea and no such continuous flow has been observed.

Along the coast of Norway there is a well-marked northward-flowing current, a more robust version of those on the Irish and Scottish west coasts. The Norwegian coastal current begins in the outflow from the Baltic Sea through the Skagerrak. The Baltic has a net input of fresh water from the many rivers in several countries along its shores. The fresh water mixes with salt water and flows out of the Skagerrak as a brackish surface layer, over a deeper salty layer that is flowing inwards. These flows in the Skagerrak maintain the water and salt balance in the Baltic Sea in the same way that the currents through the northern and southern channels do for the Irish Sea. As the outflow from the Baltic travels along the west coast of Norway, it is added to by fresh water from the coast and the water in the coastal current mixes first with North Sea and, later, Norwegian Sea water offshore.

The Norwegian coastal current can be traced as a band, about 100km (62 miles) wide and with salinity less than 35, running

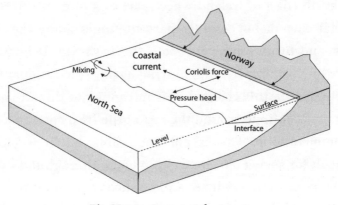

The Norwegian coastal current.

for 1,000km (620 miles) along the coast of the country. It flows in deeper water than the other coastal currents we have encountered and it is detached from the seabed. A cross-section through the flow is wedge-shaped (see picture on page 58) and the shape of the wedge can be used to work out how quickly it is moving.

The Coriolis effect is pressing the northward-moving water to its right – that is, against the Norwegian coast – and the piling up of water against the shore creates a slope in the sea surface. The slope creates a pressure head, or a pressure gradient force, which balances the Coriolis force acting on the flow. The current has come into the geostrophic balance that we encountered in Chapter 1. At the latitude of Bergen, the surface slope required to balance the Coriolis force on a current of 1m/s (for example) is about 1cm/km; the slope varies in proportion for other flow speeds. If we could measure this slope, we could work out the current directly, but it is in fact very difficult to measure such gentle slopes on the sea surface.

The water *beneath* the Norwegian coastal current is not moving. There is no Coriolis force and no need for a pressure head to balance it. Starting at the edge of the current and moving towards the coast, the sea surface rises, increasing the pressure, but because of the wedge-shape of the current a greater proportion of the water above the seabed is lighter, surface water and this reduces the pressure. The slope on the interface sets itself so that it compensates, exactly, for the change in pressure produced by the surface slope. As a result, there are no pressure variations below the interface. The change in density across the interface is small and the interface has to slope much more steeply than the surface. In the Norwegian coastal current, the gradient of the interface is about 2,000 times greater than that on the surface. Observations made with a CTD in the Norwegian coastal current tell us that the interface slopes downwards by 3 metres for every kilometre travelled shoreward. The surface slope is therefore just 1/2,000th of this, or 1.5mm/km. Using our earlier rule, this means that the flow speed of the Norwegian coastal current is about 15cm/s.

You might be wondering how the water itself does this calculation. How does the interface know that it has to slope at a certain steepness such that there is no pressure variation in the bottom water? The answer lies in a trick known to the ancient Greek philosophers and which is called *reductio ad absurdum* (reduction to absurdity). If the gradient is *not* as we have described it, there will be a residual pressure force in the bottom water and this will create currents that will move water around in the bottom until the slope of the interface is such that the pressure is equal everywhere and the currents stop.

Patience and good instrumentation are needed to detect and measure coastal currents against the background of vigorous tidal flows, which can be moving much faster. Unlike tidal and wind-driven currents on the shelf, which regularly change direction, coastal currents are following, in their unsteady way, a constant path. Averaged over time, they are imposing a circulation on our seas, a pattern in the turbulence. The currents are of limited importance for navigation; with speeds less than 1 knot, they would not trouble a motor vessel. Nevertheless, these coastal currents, driven by the steady input of rain falling on the land and turned by the Coriolis effect, are important. They transport fish larvae, algal blooms and pollutants. It is possible that fish have learned over hundreds of years to choose their spawning grounds to make use of these flows. New methods are needed to connect the pieces of coastal current that have so far been observed. Satellites may help here, as novel ways to measure salinity and sea surface height from space continue to be developed.

CHAPTER 3

A TOE IN THE WATER

ONE COLD BUT BRIGHT WINTER'S DAY, my wife and I took our youngest daughter and a flask of hot tea to a small and exquisitely lovely cove near our home. We sat on a bench above the beach, wrapped in coats and scarves, while we sipped the tea and soaked up some rare January sunshine. A noisy group of women emerged from the car park and walked, chattering and self-absorbed, past us. They paused on the beach to strip down to bathing costumes and put on brightly coloured hats and then plunged into the sea. One of them began singing at the top of her voice – perhaps as an alternative to screaming – as their bodies spread circular waves over the calm water.

Bathing in the sea at all times of year is a surprisingly popular pastime on these islands. The low temperature doesn't seem to be a deterrent. In fact, it is just the opposite: I'm told that swimming in cold water releases chemicals in the body that make you feel good. Medical research has yet to establish whether a side effect of these chemicals in some people is to make them break out into song as loud as they can. A minute spent on the internet shows that, all over the country, groups with names like the Salty Seabirds and Gullane Guillemots exist to promote swimming in the sea. The health benefits are not just physical: Rachel Ashe, the founder and director of Mental Health Swims, told me about the ways in which a dip in the sea is good for mental well-being. It makes you feel very alive and keenly aware of yourself and your surroundings.

Sea swimming is very different to bathing in a swimming pool. One obvious benefit is that the sea is free. Our islands are surrounded by an enormous open-air swimming venue, available every day of the year and with its own wave-maker. For a roof, there is a sky stretching from the shore to the horizon. You can hear the sound of seabirds calling and waves gently breaking against the shore; just like Desert Island Discs on the radio but without the theme tune. On the beach one day, I asked a friend what she enjoyed most about swimming in the sea. She paused for a moment, looked at the wide expanse of foaming water in front of us and said, 'It is the freedom.'

It has to be acknowledged that there are some features of sea swimming that won't appeal to everyone. The water is often murky: it can be difficult to see the sea floor and what might be under your feet. There are strong currents, sometimes too fast to swim against. You have to share the sea with its inhabitants, which are much better adapted to life in the water than you are. One day, a long-distance swimmer warned us to look out for seals. She had been nudged by two earlier in the day. 'They were curious rather than aggressive,' she said.

And then there is the temperature. Indoor swimming pools used for general leisure in this country are kept at a temperature close to 30 degrees C (86°F). It would be difficult to find *any* sea, anywhere in the world, consistently that warm. The temperature of the seas around our islands changes – with the seasons, with the time of day and even with the state of the tide. These changes are important to us. They control our climate, influence what we wear, the way we travel and how we live our lives. The temperature of the sea matters to the people of island nations, whether they choose to swim in it or not.

Just how cold – or, for that matter, how warm – does the sea get? A keen collector of observations of coastal sea temperatures in England and Wales is the Centre for Environment, Fisheries and Aquaculture Science (CEFAS), which is headquartered in the most easterly point on our islands at Lowestoft in Suffolk. Data is contributed to CEFAS by research and industrial organisations and by individuals who volunteer to measure the temperature of the sea, on a regular basis, just because they are interested in doing so.

Sea temperature can be measured by anyone equipped with a thermometer and a bucket. Our long-distance swimming acquaintance likes to test the temperature before she sets off and that's the way she does it. CEFAS, however, provide their volunteers with more sophisticated equipment. Today, they will most likely be given an automatically recording electronic thermometer that can store several months of data, but in the past the measurements required a more hands-on approach. Tony Scriven acted as a volunteer collector of sea temperatures for CEFAS for 18 years. Tony is now a semi-retired fisherman, based in Southwold in Suffolk. For him, collecting water temperature data fitted happily with his fishing interests. He told me that when the temperature reached 12 or 13 degrees Centigrade (54–55°F), the 'sewell' would arrive and the cod disappear (it took me a while to realise that 'sewell' was 'sole' filtered through his fine Suffolk accent). CEFAS provided him with an electronic temperature probe, a cable and a box with a digital read-out. He made the measurements from his boat in the river and tried not to miss a day: 'If I went on holiday, I'd get a mate to do it.' He was aware of the pitfalls of measuring water temperature in tidal waters and took his readings each day at the same stage of the tide. 'How cold did the water get?' I asked. 'Never below zero and usually no lower than 3 degrees,' he told me. 'It got up to 18 or 19 degrees in September, but I never saw it above 20.'

CEFAS make the data publicly available in table form. At some sites, for example Port Erin on the Isle of Man (where there was a marine station run by the University of Liverpool for many years) and Eastbourne on the south coast of England, the records are more than 100 years long and provide useful data for studying the change in our climate over that time. The average temperature in each month fluctuates from year to year but the long records are in agreement about when the sea was coldest. In January and February 1963, the average monthly sea temperature at Eastbourne (in both months) was 0.6 degrees C (33°F). That was the *average* temperature for a whole month; the actual temperature would have been lower than that on some days and (especially) some nights. In the same winter, floating ice was observed in the Menai Strait in North Wales. It takes exceptional conditions to make the sea around our islands freeze. The salt content of sea water lowers its freezing point (which is why we put salt on our roads in winter). Sea water freezes at a temperature of about -1 or -2 degrees C (30 or 28°F), depending on its salinity. There were other cold winters, in 1917 and 1947, when the mean February sea temperature at Eastbourne was 1.7 degrees C (35°F). The highest monthly mean sea temperatures at Eastbourne all occur in August. They are 19.7 degrees C (67°F) in 1899, 19.9 degrees (68°F) in 1990, 20.4 degrees (69°F) in 1995 and 19.9 degrees (68°F) in 2004. In no other year (so far) does the monthly mean temperature exceed 19.5 degrees C (67°F).

If we smooth out the temperature variation from year to year and from month to month and take a long-term average, we can see that the geographical variation in sea temperature around our islands is a small one. Although you might expect the sea temperature in the north of Scotland to be a lot less than in the Channel Islands, the difference is in fact only about 3 degrees C. The mean temperature of our waters decreases by about 0.3 degrees C for an increase in latitude of 1 degree. This trend is similar to the average for our planet; over the whole of the North Atlantic Ocean, for example, the surface temperature decreases from about 27 degrees at the Equator to about 10 degrees at 60 degrees north, a gradient of 0.28 degrees C per degree of latitude.

The decrease of sea temperature with increasing latitude is a result of our planet being spherical: the sun's rays fall more obliquely at high latitude and their warming effect is not so great. The difference in temperature from the Equator to the poles (and between Cornwall and Scotland) would be much greater than it is if we didn't have an atmosphere and an ocean working to redistribute the heat. These fluids on the surface of the Earth are continuously pumping heat poleward. Warm air and water flows north and south from the Equator, carrying heat, and is replaced by colder air and water from high latitudes travelling back towards the Equator. In the Atlantic, the work is done by the broad ocean currents on and below the surface. In our coastal seas, we don't have these major currents and other processes must be working at about the same rate to maintain the same gradient of temperature with latitude. If they weren't, there would be a sudden jump in temperature at the northern or southern edge of our continental shelf. It is likely that heat on the shelf is pumped northwards by turbulence created by the action of tides and winds.

Seasonal changes in sea temperature are caused by the tilt of the Earth's axis relative to the plane of its orbit about the sun. As the Earth moves around the sun, the solar energy received in the northern hemisphere is greatest on the summer solstice – 21 June – when the sun appears highest in the sky. Sea temperatures, however, peak in August or even in September in some places (see picture below). The lag between the summer solstice, when solar heating is at its peak, and maximum sea temperatures is ascribed to the thermal inertia of the sea – the fact that the sea takes time to respond to changes in heating. Because this lag is a ubiquitous feature around our shores, it is worth taking some time to understand how this works.

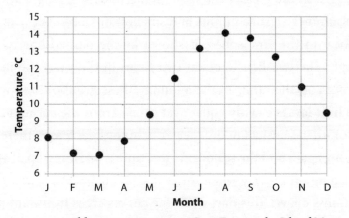

Average monthly sea temperature at Port Erin on the Isle of Man.

Heat is a form of energy and temperature is a measure of the concentration of heat. Put some heat into an iron bar or a bucket of water and the heat content, and so the temperature, increases. Most (but not all, as we shall see in a while) of the heat that enters and leaves the sea passes through the surface. Sunlight always puts heat into the sea and in our latitudes there is some sunlight every day of the year. Heat is lost from the sea to the atmosphere, and whether the sea warms or cools from one day to the next depends on the relative size of the heat gains and losses. If more heat is put in than is lost, the sea will warm up; if there is a net heat loss, the temperature will fall.

We can draw a graph of the sea's heat gains and losses against time through the year. The solar input is a heat gain with a curve like that in the picture on page 71: the value will be greatest on the summer solstice in June and smallest in December. The loss of heat to the atmosphere is shown by the dashed line on the graph. The two lines cross at two times in the year, one in the spring and the other in the autumn. These are the times when the heat losses exactly equal heat gains. From the spring to the autumn, the heat gains are greater than the heat losses; there is a net heat gain and the sea during this time is warming up. It stops warming on the day when the heat gain curve falls to cross the heat loss curve. This day, when the curves cross in the autumn, is therefore the day when the sea is at its warmest. Similarly, the other day when the curves cross, the one in the spring, is when the sea is coolest (swimmers note – this is not in December or January, but sometime later).

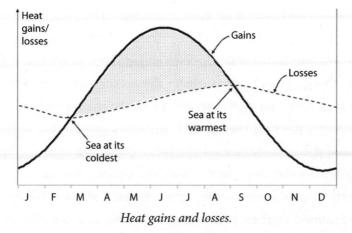

Heat gains and losses.

The heat gain curve on this diagram has a similar shape for any portion of sea on our continental shelf, although it will change a bit with latitude. The heat loss curve, however, depends on how the temperature of the sea surface changes as heat is put in or taken out. Heat losses from the sea to the atmosphere, generally, increase with the temperature of the water surface. This is why the 'loss' curve peaks when the sea is warmest. The *shape* of the curve depends on the water depth. A shallow water body will warm quickly; the heat loss curve will rise rapidly and meet the heat gain curve early in the year. The maximum temperature in shallow water may therefore occur in July or early August. In deep water, especially if the heat is mixed downwards so that the surface stays cool, the rate of increase of temperature during the summer will be slower. The heat loss curve will rise more slowly and the maximum temperature will be later in the year: perhaps in late August or early September.

And so we learn an interesting and important point: the time of year at which the sea will be warmest depends on how deep and well-mixed the water is. We can learn one other useful fact from this diagram before we leave this section. The area between the heat gain and heat loss curves in the spring and summer represents the heat that is, each summer, *stored* in the sea. I've shaded the heat storage on the picture and you can see that it depends on the shape of the heat loss curve. More heat will be stored in mixed, deep water, which keeps a cool surface during the summer. The heat is then released in winter to keep us warm.

One of the first people to recognise the proper relationship between heat and temperature was the Scottish chemist and medical practitioner Joseph Black (1728–99). Black held the chair of chemistry first at Glasgow (where he did most of his research) and later at Edinburgh (where he became a renowned teacher). The chemistry departments at both these universities are named after him. Black carried out experiments on the heating of different materials and established the idea of heat capacity: the quantity of heat needed to raise the temperature of an object by one degree. He would take an insulated flask containing hot water (of known temperature) and place into it an item made of the material whose heat capacity he wished to measure. For instance, if he dropped in a gold sovereign, initially at room temperature, the temperature of the water in the flask would fall a little (and the temperature of the sovereign increase) as heat was transferred from the water to the gold. Material with a high heat capacity would produce a large drop in temperature as it sucked a lot of heat out of the water and vice versa.

Black also discovered the idea of latent heat. He noticed that when water in a saucepan placed over a flame boils, its temperature remains at 100 degrees Centigrade (212°F) even though it is still taking in heat. He surmised, correctly, that the heat was being used to change the water into steam: energy was needed to change the state of the water from a liquid to a gas. He anticipated, also correctly, that energy would be needed to change ice into water and he waited eagerly for a cold winter to descend on Glasgow so that he could test out his ideas.

Black's measurements of the heat capacity of materials were all relative to that of water. He could tell you that the heat capacity of gold was 3 per cent that of water, but not what that meant in absolute terms that had a meaning outside the laboratory. That advance in understanding had to wait until the experiments of James Joule (1818–89). Joule was born into a brewing family in Salford, Lancashire. The family business survives today, 200 years after his birth, although it has moved its base to Market Drayton in Shropshire. Joule measured the heat capacity of water in terms of the fundamental units of mechanical energy in a series of remarkable experiments, made while he was also managing the brewery. He warmed water by stirring it with a paddle, which was driven by a falling weight. He could measure the energy put into the water by the distance the weight fell. His idea was that the potential energy in the falling weight was first turned into kinetic energy as the water swirled around in turbulent eddies. Within the turbulence, friction – or viscosity – between individual pieces of water rubbing against each other would convert the kinetic energy into heat.

It is tempting to think of Joule sampling the products from his brewery as he carried out his experiments, but that is unlikely. The temperature changes involved were very small and Joule had to guard against heat escaping through the sides of the container. His experiments demanded great skill and care. He needed to keep a clear head.

Not many gravestones carry a number other than relevant dates, but James Joule's does. The number is 772.55, which was his best shot at the amount of energy required to raise the temperature of one pound of water by one degree. He measured energy in units of foot-pounds (the energy released when a weight of one pound fell a distance of one foot) and temperature in degrees Fahrenheit. In today's units, which bear his name, his figure is equivalent to 4,093 joules per kilogram of water per degree Centigrade. The modern value, in round numbers, is 4,200 joules per kilogram per degree Centigrade.

Joule also, famously, took a thermometer on his honeymoon so that he could measure the change in water temperature from the top to bottom of a waterfall near Chamonix in the French Alps. He wanted to see how much of the potential energy released as the water fell was turned into heat. Some energy will be converted into spray and the sound in the noise of the waterfall, but much will become turbulence as the water churns around at the bottom of the fall. Within the turbulence, energy would be converted into heat in the same way as it was in his experiments in Lancashire. One kilogram of water falling 30 metres will release about 300 joules and if all that energy is turned into heat, the temperature of the water will rise by about 0.07 degrees Centigrade. Measuring such a small change of temperature in the open air challenged Joule's technical capacity and his experiment failed on this occasion.

A modern version of Joule's measurement of the heat capacity of water is to boil a kettle. The kettle in our home is rated, according to the label on the bottom, as 2,500–3,000 watts. The watt, named after another pioneer of the Industrial Revolution, is a rate of energy input or output of 1 joule per second (I'm not sure why there is a range, the kettle is a fancy one that is supposed to be economical and perhaps it is something to do with that). It takes 2 minutes and 3 seconds for the kettle to boil 1 litre of water. The energy put into the water in that time is 369,000 joules (if we take the high end of the rating and multiply it by the number of seconds). The water started at a temperature of 15 degrees C and so the increase in temperature up to boiling is 85 degrees C. This gives an estimate of the heat capacity of water as 369,000/85 = 4,341 joules per kilogram per degree Centigrade (since 1 litre of water has a mass of 1kg). Bearing in mind that my kettle is not very well insulated, that's not bad.

Water has an unusually high heat capacity. It is not the highest of any known substance, nor even the highest among liquids, but it is high among materials that are commonly found on the surface of the Earth. In particular, the heat capacity of water is higher – by a factor of three or four – than that of materials that make up the land. During summer heating, the sea warms up more slowly than the land and, because it keeps a cool surface temperature, it is very good at storing the summer heat. The storage is greatly enhanced by the fact that water is a fluid and can move around. The part of the land that warms up during the summer is limited to just a few metres downwards from the surface. In the sea, parcels of surface water can be mixed downwards tens or hundreds of metres below the surface, carrying their heat with them.

The storing of large amounts of heat in the sea and its subsequent release tempers the climate of coastal communities. Towns and villages by the sea experience smaller yearly and daily variations in temperature than those inland. This effect is measurable even on our small islands. The average monthly temperature at Oxford in the English Midlands, for example, varies from 3.5 degrees C (38.3°F) in January to 16.9 degrees C (62.4°F) in July – a range of 13.4 degrees (24.1°F). On the Scilly Isles, the range is from 7.3 degrees (45.1°F) in February to 16.3 degrees (61.2°F) in August – a change of just 9 degrees (16.1°F). The variation in air temperature from day to night is also smaller at the coast, where night-time temperatures tend not to drop as low as those inland as heat is released from the sea. Frosts are also rarer in coastal areas. These changes within our islands are small, however, compared to the difference between the climate of all these islands and that of a continental interior. We are spared the bitterly cold winters and torrid hot summers of, say, Eastern Europe. Everyone in Britain and Ireland benefits from the proximity of the sea.

The tide could also play a role in influencing the variations in air temperature at the coast. If high tide happens to be at midnight on a cold winter's night, the heat stored in the sea will help to keep things warm and prevent a frost. Seven days later at the same place it will be low tide at midnight and the warming effect will be much less. We might see a fortnightly variation in winter night-time air temperature controlled by the phasing of the tide. A similar thing should happen in summer but now a high tide in the middle of the day will serve to keep things cool. These regular signals in air temperature will be superimposed on the more random variations associated with passing weather systems. As far as I know, these effects have not been formally reported, but I think they would be fun to look out for.

The heat entering the sea in spring and summer warms the water at the surface. The warm water is buoyant; its density is lower than cold water that lies beneath it. The sea becomes *stratified*, with a warm surface layer lying on top of cold water. Winds blowing over the surface mix the sun's heat down to some extent but there is a limit to how far they can do this. A sudden squall may push some of the surface water down into the colder deep layer, but it will then be pushed back by its buoyancy. The sun's heat continues to go into the surface layer (with only a little leaking through to the bottom), making the surface warmer and more buoyant as summer progresses; the increasing buoyancy of the surface water makes it even harder to mix the heat downwards. A sharp interface develops, separating the surface layer from cold bottom water. In the late autumn, the net heating rate becomes negative and the surface begins to cool. The density difference between the surface and bottom reduces. The weakening stratification is more easily disrupted by the increasingly strong winds at this time and, in the last months of the year, the seas on the shelf become completely mixed from surface to bottom.

The building up and breaking down of stratification with the seasons is important to the biology of shelf seas. At the base of the food chain in the ocean are the tiny photosynthetic organisms called phytoplankton. These creatures are the 'grass of the ocean' but, unlike terrestrial grass, they have no roots. Instead, they are free to drift with the currents (the word *plankton* comes from the Greek for drifter). A drop of coastal water viewed under a microscope may contain a number of different phytoplankton cells. There is a bewildering variety of types, each one adapted to take advantage of some aspects of living in the sea. In order to grow and multiply, phytoplankton need water (which they have in abundance), light (for photosynthesis) and certain nutrient elements such as nitrogen, phosphorus and silicon, which are used to build their cell structures. The collective photosynthesis of ocean phytoplankton produces about half the oxygen in our atmosphere.

When the shelf seas first stratify in spring, phytoplankton living in the surface layer have plenty of light and – to start with – plenty of nutrients too. They grow rapidly in a spurt, called the spring bloom, which lasts until the nutrients in the surface layer are used up. Grazing by tiny animals creates a shower of pieces of plankton, which fall down to the seabed, carrying their nutrients with them. In the deep ocean (which, for a lack of mixing energy, is permanently stratified), there is a continuous drain of nutrients away from the surface. The growth of phytoplankton in the surface ocean is limited by the poor supply of nutrients. In shelf seas, however, the nutrients that have fallen to the seabed are released back into the entire water column when the sea becomes completely mixed in winter. They are available to fuel the next spring bloom when the water stratifies in the following year. Shelf seas are measurably more productive, biologically, than the deep ocean and it is likely that the ready recycling of nutrients on the shelf is, at least partly, the reason.

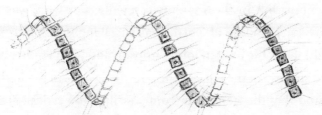

A helical chain of phytoplankton cells.

Not everywhere on the shelf stratifies in the summer. Tides are vigorous mixers of everything in the sea, including the sun's heat. In places where tidal currents are strong – the southern North Sea, most of the Irish Sea, the English Channel and around islands everywhere – stratification does not happen at all. Instead, the turbulence generated by the tides and winds together has enough power to mix the sun's heat continually down to the seabed, creating vertically mixed conditions throughout the year. This can happen even in the deepest water on the shelf. The North Channel of the Irish Sea, for instance, is 200 metres deep but the strong tidal flows here are able to stop stratification developing in the summer. Because of the vertical mixing, the North Channel is particularly effective at storing the sun's heat and the adjacent land – the Mull of Galloway and the eastern coast of Northern Ireland – is usually favoured with mild winters.

In the summer months, the seas around our islands are divided into two distinct regimes – those that are stratified and those that are not. This much was known in the early part of the 20th century but in the 1970s two scientists at what was then the University College of North Wales in Bangor began to think seriously about the matter. John Simpson and John Hunter made detailed surveys of the edge of the stratified water in the Irish Sea and found that the transition from stratified to vertically mixed water was remarkably sharp. They called this boundary a tidal mixing *front*. Surface temperatures on the stratified side of the front were warmer, by a few degrees, than those on the mixed side and the transition from one to the other occurred in just a few kilometres. On calm days there was sometimes a foam line, which suggested surface currents were converging on the front, and a keen eye could spot a change in the colour of the water.

Simpson and Hunter applied a neat energy argument to predict where these fronts would occur, not just in the Irish Sea but all around our islands and, indeed, in temperate seas worldwide. Exactly at a front, they reckoned, the tide has just enough energy (no more, no less) to mix the sun's heat down to the seabed. The energy needed for the mixing is proportional to the water depth and the energy input by the tide is proportional to the cube of the current speed. Fronts should occur, therefore, at places where the water depth was equal to a fixed number multiplied by the cube of the tidal current speed. Because depth contours and the distribution of tidal current speeds are geographically fixed, tidal mixing fronts (unlike fronts in the atmosphere) would always occur in the same place. This idea could be tested by observation. Making a prediction in science, as in other fields, is always risky. Failure can affect the rest of your career. Simpson and Hunter's prediction was a bold one, but it was also timely. There was about to be a change that would revolutionise the way we could observe the ocean and which would provide a rapid and conclusive test of their theory.

In the 1970s, new satellites were being put into orbit about the Earth to provide data to help with the growing science of weather forecasting. The satellites were equipped with conventional cameras to photograph cloud patterns and also infrared cameras, which could measure the temperature of an object. From its vantage point several hundred miles above us, a satellite could quickly map the temperature of large sea areas, provided clouds didn't get in the way.

In the UK at that time, satellite data (mostly from the United States NOAA series of weather satellites) was received at the University of Dundee. The engineers who set up the station, John Brush and Peter Baylis, had – to begin with, at least – no official remit (although Dundee did later become the official receiving station for the UK). Baylis and Brush weren't directly interested in the pictures they made; they set up the station because they were gifted engineers who enjoyed the technical challenge of the problem of receiving and processing the signals from the satellite. They did such a good job, and their products were of such high quality, that the pictures were soon in demand for newspaper and TV weather bulletins. The station welcomed scientists from the UK and Europe who were interested in using their pictures for research, and engineers from all over the world who were keen on learning how to set up their own receiving station.

One group of visitors to Dundee who had a particular interest in the pictures was from the Royal Navy. The speed of sound in water depends on the water temperature and, when the water is stratified, sound rays are refracted at the thermocline – the horizontal layer separating the warm water above from cold water below – in the same way that a light beam is refracted at the interface between water and air. Submarines could hide beneath the thermocline and avoid detection by sonar. Infrared satellite images were a powerful tool for detecting where this could happen and this fact was of great interest to the navy. In later years, the satellite images were supplemented with computer models of the subsurface indicating the depth of the thermocline and the temperature change across it.

Towards the end of the 1970s I joined the stream of visitors to the satellite receiving station. The city of Dundee was approached, then as now, by trains from the south, across the 4.5km (2³/₄ miles) of the Tay Railway Bridge. The bridge curves as it approaches the north shore of the river and travellers get a good view of the bridge and their train crossing it. From the station it is a brisk step, past the city centre shops selling Dundee cake, up the Perth Road to the university. The receiving station was housed in the Electronic Engineering department (although I have a feeling that the large satellite receiving dish was placed on the roof of an adjacent university building, which was taller and had better reception). Research laboratories in physics and electronics departments looked different in the 1970s. This wasn't entirely down to the flared trousers and other fashions of the day; the art of miniaturising electronics had not yet been perfected. A piece of circuitry that could today be put on a microchip then required a box the size of a refrigerator. Computers for use in a laboratory were still in their infancy and there were few desktop and no laptop computers.

The receiving station was crammed with metal shelves filled with circuit boards, coloured wiring, flashing lights, knobs, dials, oscilloscopes and large tape recorders for storing data. As they showed visitors around, Baylis and Brush would sometimes pause to make adjustments to instrument settings. The whole thing was a fine example of pressing what was available into service and building the rest yourself. And the results were absolutely fantastic.

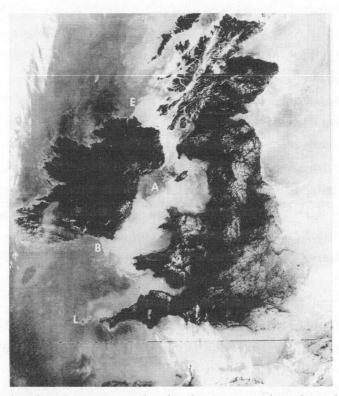

Infrared satellite picture produced at the University of Dundee in the early 1980s. The letters in white mark the positions of tidal mixing fronts.

The satellite pictures were printed on high-quality photo-facsimile machines donated by a local newspaper office (the original purpose of these machines was for sending news photos from one office to another down telephone lines). I've shown one of the pictures on page 84. They were printed in black and white on glossy photographic paper, 10 inches on a side. Land was shaded black and clouds white and the sea in various greys, dark grey being warmer than light grey. You could hold a picture in your hand, fresh from a satellite pass that morning, and see (through the gaps in the clouds) the surface temperature of the seas around our islands from the Bay of Biscay in the south to the Shetland Islands in the north and from the shelf edge in the west to the coast of Norway in the east. And there, exactly where Simpson and Hunter had predicted, lay the tidal mixing fronts: caught as changes in the greyness of photographic emulsion. A satellite had done a job in a few minutes that would have taken a research vessel several months.

Satellite data, like ship data, has its limits. Infrared images tell you only what is going on at the sea surface, but they soon became part of the toolbox available to oceanographers. Importantly, they provide a synoptic picture, an instantaneous snapshot of sea temperatures over a large area. The satellites showed the formation and break-up of stratification in our waters. Starting in the spring, the area of stratified water quickly spreads over the shelf from the shelf break to the locations where strong tides stop it advancing further. It retreats much more slowly in the autumn because it takes a while to destroy the buoyancy stored in the stratified water. During the summer, the edge of the stratification moves back and forwards a little with a fortnightly cycle as enhanced tidal mixing on a spring tide nibbles away at the edge of the stratified water, which then re-forms on the subsequent neap tide.

Stratification on the shelf finally disappears in some places as late as December. As the winter storms destroy the last vestiges of the summer stratification, heat is mixed down to the bottom of the sea. At these places (south of Ireland is one example), the seasons at the seabed are almost the reverse of what we experience on land. During the summer, the bottom water remains cold as the sun's heat is trapped in the surface layer. It is only in the winter, when the stratification is finally destroyed, that the surface warmth finally makes it down to the seabed. The animals that inhabit the muddy bottom of our deepest and most poorly mixed waters experience their highest temperatures of the year around about Christmas.

The temperature of our seas can mostly be predicted very well from the net heat flux that passes through the water surface. The sea gains heat from the sun and gives it back to the atmosphere through a mixture of processes – conduction, evaporation and radiation – which are all fairly well understood and predictable from meteorological observations. There are, however, at some times and in some places, discrepancies between the observed temperature of the sea and that calculated from the heat flux through the sea surface using the best available meteorological data. One reason for this discrepancy is that heat can be carried horizontally by the (generally weak) currents on the shelf.

Remarkably, this idea can be turned on its head. We can use the *difference* between maps of calculated and observed sea temperatures to estimate the speed and direction of the circulating currents on the shelf. The person who had the confidence to try out this idea in the shelf seas of north-west Europe was Alan Elliott. In the picture on page 88, I show a map of the flows that he estimated in this way. The calculated current velocities are small, less than 10cm/s, but they agree with what we know about the residual flows from other methods. There is a current moving northward between the west coast of Ireland and the shelf break, which continues along the west coast of Scotland at a speed of between 5 and 10 cm/s (here, it is the Scottish coastal current that we met in Chapter 2). The current then flows down the western side of the North Sea to the latitude of the Dogger Bank, where it turns to the east towards Denmark and Norway. Here, some of the flow enters the Skagerrak and some turns north to join the Norwegian coastal current. This is an example of using a tracer – in this case temperature – to form a picture of the circulation on our shelf. When the circulation is slow and variable and hidden behind much faster tidal currents, using a tracer is a good way to do this job.

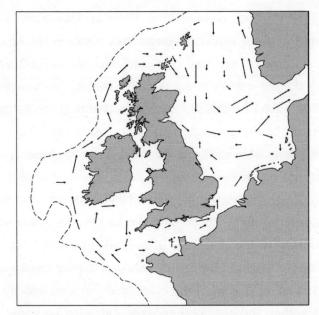

*The circulation on the shelf inferred from discrepancies in sea
temperature. The longest arrows shown represent currents
of about 5 cm/s.*

Are the sea temperatures around these islands changing with
global warming? For sure they are. Two of the longest data sets in
the CEFAS records are at Port Erin on the Isle of Man (since 1903)
and Eastbourne in East Sussex (since 1892). At both places, over
the length of the available data, the annual mean temperature has
been increasing at a rate of just under 0.1 degrees C per decade.
The trend is accelerating. If data from just the last 25 years is
used, the rate of increase is 0.4 degrees C per decade.

Longer records of sea temperature on the shelf, going back hundreds or even thousands of years, can be created by analysing the growth lines in the shells of certain marine molluscs. One particularly long-lived mollusc has the scientific name *Arctica islandica*. It lives on the sea floor in the sediments of the continental shelf and each year adds a new line, or growth band, to its shell. The age of these animals can be determined by counting the lines, in the same way that we count rings to age a tree. One *Arctica* individual was found to be over 500 years old; it was nicknamed 'Ming the mollusc' because it was born during that particular Chinese dynasty. These are probably the longest-lived animals on our planet, although why they should want to live so long is a mystery to me; their lives must be exceedingly boring.

These unassuming animals have, however, provided a rich source of information about past sea temperatures. Material extracted from the growth bands can be analysed for the ratio of oxygen isotopes and this gives an estimate of the temperature of the water in that year. It is not necessary to collect only living animals to do this. The width of the growth bands varies from year to year; there is a pattern to them like the lines on a barcode. Animals whose lives have overlapped will have matching patterns in the period of the overlap. Starting with a live animal, the early part of the pattern can be matched to the later part of the pattern on the shell of an older, dead, animal and the age of that shell can be established. This, in turn, can be matched to the pattern on an even older shell and so on. In this way, long records of sea

Arctica islandica.

temperature can be created. Sea temperature records established in this way match known past fluctuations in climate including the little ice age of the 17th century, when ice skating was possible on the Thames.

You can sometimes find *Arctica* shells on the beach. When the animal is alive, it lives inside two shells, hinged at their base. When the animal dies, the hinge breaks and the shells are separated. The two parts look similar, but are in fact mirror images of each other – like your left and right hand. The individual shells are particularly large and handsome specimens, about 10cm (4in) across, and there is a curious thing about them. The discoverer of this curiosity, Paul Butler, took me to Newborough beach on Anglesey to show me. Newborough is a long beach facing the prevailing south-westerly winds. Starting at one end of the beach, most of the *Arctica* shells are left-handed.

In the middle, left-handed and right-handed shells are present in equal proportions and at the far end of the beach, they are mostly right-handed. There is some sorting mechanism operating on the shells.

I doubt anyone knows for sure how this sorting mechanism works, but it is likely to do with the way the shells are carried by moving water. We can imagine that there is a colony of *Arctica* living offshore. When an animal dies, its two shells are carried towards the shore by currents driven by the prevailing winds. As they move, the left-handed shells veer one way and the right-handed shells veer the other in response to the way the current acts on the particular shapes of the left- and right-handed shells. Paul and I tested this idea by putting some shells in a flowing stream; as they were carried along, one half of the shell moved towards one bank and the other to the opposite bank. It's a plausible hypothesis, but it is a long way from certainty. The sorting of shells in this way, though, is something to look out for on your next walk along a beach. Paul tells me that the beach should be long and the shells that you choose to look for should be large and heavy.

CHAPTER 4

WAVES ON A CORNISH BEACH

M OST PEOPLE WHO LIVE IN THESE islands have stood on a beach and watched waves coming towards the shore from the sea. In general, these waves are of two types. There are short, choppy ones, which have been generated by local winds and have not travelled very far. There may also be regular long-crested waves, the sort loved by surfers. These swell waves have been created in a storm in the deep ocean and have travelled a long way to get to the beach.

Watching waves being made by the wind can be fascinating and you can see things that you don't expect. It's best if you have some vantage point, such as a cliff overlooking a small bay or lake. On a visit to Penzance in Cornwall, I took a short walk along the coast path to Newlyn, where the mean sea level used as the basis for Ordnance Survey maps is measured. Newlyn is a busy and prosperous fishing village with a row of fishmongers facing the sea. At one point the road rises to give a good view of the fishing harbour and I stood watching a light breeze making small waves on the water surface. Much of the surface was calm, but the wind made patches of waves that travelled over the surface of the water in the wind direction. I tried focusing on a single wave crest in the patch but I couldn't follow it for long before it disappeared. I would then pick up another crest and the same thing happened. Although the patch of water with waves in it was moving across the harbour, it wasn't possible to follow a single wave in the group of waves for very long.

This is the sort of observation that makes you wonder if you are seeing things properly. It's not what you'd expect. I would expect a wave crest to travel gracefully across the water until it hit an obstacle, but it wasn't like that: these crests were really not travelling very far before they disappeared. It is, in fact, a well-known feature of the theory of waves travelling in water that is deep compared to the distance between their crests (a distance known as the wavelength), but it's one thing to be familiar with the theory and quite another to see it in action. I recommend that you look out for it next time you are by the sea.

It is probably fair to say that the exact way that wind makes waves on a water surface isn't understood properly – not, that is, to the extent that accurate predictions can be made of the sort of waves that will be created in any conditions. In general, though, the way that the wind ruffles up the surface of the sea into waves goes something like this. The air above the sea surface is turbulent: there are small variations in the pressure of the atmosphere pressing down on the sea and these create small undulations in the water surface. The surface is depressed a bit where the air pressure is high and, a short distance away, the surface is raised a little where the pressure is low. The size of these undulations created by variations in air pressure alone is not very large, but they provide the key for a horizontal wind blowing over the sea to make waves. Imagine what happens as the wind blows over one of these upward perturbations in the sea surface. On the upward slope, the one facing into the wind, the wind pushes water in the direction it is blowing – that is, towards the top of the slope. As the wind enters the sheltered, or leeward side of the nascent wave, it forms an eddy. It curls round under itself, blowing back towards the top of the wave on the leeward side (see picture on page 96). You can often feel this eddying motion in the wind when you hide behind a bus shelter in a storm. The wind blows over the top of the shelter and then eddies around to come at you in exactly the opposite direction to the wind before it reaches the shelter. I'm told that small sailing boats that find themselves in the lee of a large wave in the ocean can feel this reversal of wind direction and are sometimes 'taken aback'.

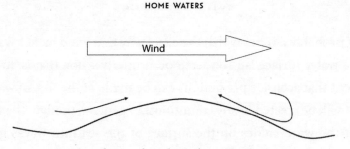

How the wind might make a wave grow.

Anyway, the thing is that, if there is a small hump in the sea surface sticking up into the wind, the wind in contact with the water blows towards the top of the hump from both the side facing into the wind and the side facing away from the wind. Water will be carried from both sides towards the top and the hump will grow into a wave. This process will continue as long as the wind continues to blow over that bit of the sea surface or until the wave reaches such a height that it loses water by tumbling over itself as quickly as the wind pushes water up towards the crest of the wave.

That's the way the wind was making the small waves in Newlyn harbour while I was watching and it is essentially the way that the wind makes storm waves in the deep ocean, although other things might be important there too. The waves move along with the wind, taking energy out of the atmosphere and using it to build crests and hollows in the sea surface. If the wind were to stop, or the waves moved into a sheltered spot, the waves would carry on with a life of their own, until they lose their energy through internal friction in the water or by breaking on the shore. This independent existence isn't possible for all types of waves – for example, the waves created in a field of wheat stop moving soon after the wind ceases to blow.

The most obvious control on the size of the wind waves made on the sea is the strength of the wind, an idea incorporated in the Beaufort scale used to classify sea states. The attraction of the scale is that it relates the wind strength to the sea state in ways that can easily be observed, and are important, at sea. The Beaufort scale runs from 0 (calm) up to (usually) 12 (hurricane force) and the wave height increases in an upward curve with the scale number. When I was watching the waves forming in Newlyn harbour, I made a note that I could feel the wind on my face, so that would have been a light breeze, force 2, which should have been making waves of a height between 30 and 60cm (1–2 feet). I would have said that the waves I was watching were smaller than that – certainly not 60cm – but I was at some height above the water (wind speed tends to increase with height above the sea) and the waters of the bay were sheltered. It is also likely that the waves didn't have the room – or the time – to grow to their full size.

The height of sea waves also depends on the distance, called the *fetch*, that the wind has to act upon the sea. If you've ever watched the wind whipping up waves on a small lake – the sort you might see in a park – you may have noticed that the waves are biggest on the weather shore, the side of the lake towards which the wind is blowing. Energy is extracted from the wind to make waves and, as the fetch increases, more energy can be transferred and the waves grow. A good working relationship between wave height and fetch was established by Thomas Stevenson (1818–1887), an engineer and the father of the author Robert Louis Stevenson. Stevenson senior built marine structures – lighthouses and harbours – and so was interested in sea waves and the forces they bring to bear on objects that get in their way. The rule he came up with, through careful observation, was that the height of the waves increases as the square root of the fetch. This is a sound idea physically. The work, or energy, imparted by any force acting on an object increases in direct proportion to the distance the force moves: in the case of the wind blowing over the sea, this distance is the fetch. The energy in waves is proportional to the *square* of the wave height and so, balancing energy input to energy acquired, it must be that wave height depends on the square root of the fetch. It's a neat idea. According to Stevenson's rule, the wave height in metres is equal to one-third of the square root of the fetch, measured in kilometres. The largest waves that could be made in a pond 100 metres across would be 11cm high, no matter how strong the wind blew, and this was probably the factor limiting the waves I saw in Newlyn harbour. In a sea with a possible fetch of 100km, waves could grow to 3m in height and in an ocean 4,000km across, the limiting wave height would be about 21 metres.

One of the most prolific early observers of waves in the sea (and all sorts of other places, including desert sand) was Vaughan Cornish

(1862–1942). Cornish was what we might call a gentleman scientist: a man of independent means who delighted in spending his summer holidays in a howling gale in the Bay of Biscay just so that he could get a good look at the waves. Cornish wrote about his observations in a book *Ocean Waves and Kindred Geophysical Phenomena*, published in 1934. He was an exceptional observer, noticing things many people would miss, and he had the ability to describe the simple measurements he made in a vivid and inspirational way. Anyone reading Cornish's book could dream that they too could go out and learn new things about the natural world, and quite rightly so. Here he is in full flow on those waves in the Bay of Biscay, as ships struggled with the storm all around him:

At 8 a.m. as I stood on the promenade deck, with an eye-height 27 feet above the water-line, each passing wave was well above the horizon. From the look of the sea I judged the waves to be as high as any which I had seen on previous voyages, that is to say, 40 feet from trough to crest. They were remarkably uniform in height and were much steeper in front than at the back, thus differing from the almost symmetrical swell of the previous day. I timed the arrival of successive wave crests and found that the 'period', or interval between them, was 13.5 seconds. The theoretical wavelength for this period is 934 feet.

Cornish was using a theoretical relationship between the period of waves (the time interval between one crest and

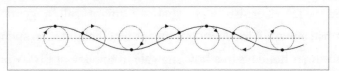

The circular motion of water particles in a wave in deep water.

another) and the wavelength (the horizontal distance between crests). The background to the theory goes like this. The humps and hollows of water made on the surface by the action of the wind cannot be sustained without some kind of movement. Gravity, acting on a piece of the sea surface that is raised above the mean level, pulls it down and the water below moves sideways to make room for it. To preserve the total volume of the water, the surface is then pushed upwards in another place. This is how the shape of the wave – its highs and lows – moves along.

The kind of movement, or water currents, created by the wave depends on how deep the water is. In water that is very deep compared to the wavelength, the currents go round in circles (see sketch above). If we follow a small parcel of water as the wave passes, it is at the top of its circular motion at the crest of the wave and at the bottom of the circle in the hollow, or trough, of the wave. Halfway between the crest and the trough, the parcel is halfway between the top and bottom of its circular motion. As the parcels of water move around their circles, the wavy surface travels along above them. In the picture, the wave is moving from left to right and if, for example, we look at the second circle from the left we can anticipate that, as the black dot moves clockwise, the water surface rises to meet the crest arriving from the left. You can see the circular motion of the water in real waves by tossing something that floats on to the sea surface where waves are passing. A cork makes a good marker for this experiment.

As the wave passes, the parcels of water keep their mean position fixed (more or less): the water doesn't move along very much with the wave. As we go deeper beneath the wave, the size of the circle and the currents get smaller. Submarines can dive to avoid the worst of the motion of storm waves. As the waves move into shallow water, the circles are squashed into ellipses and in very shallow water compared to the length of the wave, the currents just move horizontally: forwards under the crest of the wave and backwards under the trough. In deep water waves, where the motion is circular, the fact that the currents get less as you go down from the surface means that the speed at the top of the circle is a little faster than the speed at the bottom of the circle. During a complete loop around the circle there is therefore a net movement in the direction the wave is travelling. This residual motion created by waves is called *Stokes drift*, after the Irish physicist Sir George Gabriel Stokes (1819–1903). It is usually small, less than 1 knot, but needs to be taken into account sometimes, for example when planning search and rescue missions at sea.

Anything that moves in a circle needs a centripetal force, acting towards the centre of the circle, to maintain the motion. When a parcel of surface water is halfway between the top and bottom of its circular motion, this force is provided exactly by the horizontal pressure gradient force created by the slope of the water surface. In turn, the slope of the surface depends on the height and length of the wave. That information is all we need to work out the relationship between the wavelength and the time it takes for a water parcel to complete a circle (which is the same as the period of the wave). The solution tells us that the length of a wave in deep water (in metres) is equal to 1.56 times the square of the period in seconds. Wave crests that arrive, say, every five seconds, must have a wavelength – a distance between crests – of 39 metres (1.56 times 5^2). Ten-second waves in deep water have a length of 156 metres. The equivalent formula in imperial units is that the wavelength in feet is equal to 5.13 times the square of the period in seconds and that is the relationship Cornish used in the quote above (5.13 times $13.5^2 = 934$).

The speed at which the crest of a wave travels over the ground is equal to the wavelength divided by the period. The speed of waves in water that is deep compared to their wavelength is therefore 1.56 times the period, if the period is measured in seconds and the speed in metres per second. Waves with a longer period (and so also a longer wavelength) travel faster than short-period waves. I saw a nice illustration of this when

I worked on the shores of the Menai Strait in North Wales. Some of my colleagues were housed in offices on an island in the strait, connected by a stone causeway to the main island of Anglesey. Occasionally, a speedboat would be whizzing along the strait as I was crossing the causeway and, shortly after the boat had passed by, its wake would start lapping on the shores of the island. When I knew what to look out for, I would listen to the sounds made by the waves in the wake as they arrived and, sure enough, the period of the lapping would steadily decrease as I walked across the causeway. The long, faster waves in the wake of the speedboat had arrived first followed by the slower, shorter ones.

The fact that different wavelengths travel at different speeds explains why it was difficult for me to follow a single wave crest for very long in the harbour in Newlyn. The wind was making waves with lots of different wavelengths (the atmospheric turbulence that was creating the initial disturbances in the sea surface has no preferred length scale). The different lengths of wave in the patch of roughened water that I was watching travelled at different speeds and interfered with each other to create the overall pattern of the sea surface at any one time. Sometimes the crest in one wave would coincide with the crest of another and they would add together to create a particularly large crest that would catch the eye. But this wouldn't last for long. The wave with the longer wavelength would gradually move ahead of the one with a shorter wavelength and soon the trough of one wave would be in the same place as the crest of the other. The two waves would then interfere destructively and what was a large crest a few seconds earlier would now be a small crest, or no crest at all.

A group of waves made by the wind in a patch of sea water behaves in a particular way. Each wave in the group travels at a speed that depends just on its period, or wavelength, as we described above. This speed is sometimes called the *phase* speed of the waves. However, the biggest waves in the group, where the crests of different waves overlap, travel at a different speed. The position at which the individual waves overlap to make the biggest wave travels at a somewhat slower pace than the waves that create it: the speed of the highest wave is called the *group* velocity. It can be shown with some maths that the group velocity is just half the phase speed. So, a group of waves with a period centred around 10 seconds, for example, will have a phase speed of 1.56 times 10, namely 15.6m/s, or about 35 miles per hour, but the point with the biggest wave – the centre of the group – will travel across the water surface at a speed of only 7.8m/s or 17.5 miles per hour.

It can also be shown with some maths that the energy in the group of waves travels at the group velocity. The packet of waves sticks to where the energy is and so travels across the ocean at the group velocity. An individual wave, with a particular wavelength, passes through the group, reaches the front of the group and then *dies away* as it moves out of the region that contains the energy (a new wave with this wavelength grows at the back of the group). The patch of waves moves sedately across the sea at the group velocity, the speed of the energy, and individual waves move through the patch, growing as they move from the back to the middle and then shrinking again as they move to the front of the patch. This strange behaviour of a group of waves only happens when different wavelengths travel at different speeds. As the waves move into water that is shallow compared to the wavelength, the speeds of the different wavelengths all become the same (and just depend on the water depth). We no longer see this behaviour. That is why we don't see waves appearing and disappearing when we watch waves approaching a beach: they are travelling in water that is shallow compared to their wavelength.

In the 1940s, there was a renaissance in interest in the way that ocean waves behave. It was important to know about waves because troops and equipment were being landed on beaches and the success of the operation could well depend on the sea conditions at the time. Local weather forecasts were important but not enough on their own: large waves, created by a distant storm, could arrive at a beach on what was a calm day locally and create havoc with landing craft. Military operations that involved amphibious landings needed predictions of the likely wave conditions they would encounter on the landing beach, including swell arriving from distant storms. This need grew more acute as the Normandy landings planned for June 1944 approached.

The scientists who were given the task of wave forecasting for the D-Day landings didn't have much background to go on. The theoretical work that was of most help was over 100 years old, written by the French scientists Augustin-Louis Cauchy and Simeon Denis Poisson. New methods were needed and these had to be built up from scratch. The starting point for the forecast was a weather map of the North Atlantic, created using observations from weather stations and ships at sea. Intense low-pressure systems on these maps identified storms capable of generating swell waves that could travel great distances. Winds would spin around the low-pressure systems and it was when the wind was blowing towards the beach of interest that it would generate swell that would travel in that direction. The spacing of the isobars gave the wind speed and empirical rules were used to estimate the height of the waves that might be generated by these winds. It was assumed that, while the waves were being generated, the fastest waves would travel no quicker than the wind speed. If they did, they would outstrip the wind and not be able to extract so much energy from it. This assumption allowed the longest wave period to be calculated using the fact that the phase speed in metres per second is 1.56 times the wave period in seconds. For example, wind speeds of 50 knots, or 25 metres per second, could be reckoned to generate waves with a period of $25/1.56 = 16$ seconds (as well as waves with shorter periods). That was enough information to work out the group velocity and so to estimate the time it would take for the waves to travel from the storm to the beach. Since it would likely take a few days for the swell waves to reach Europe from a mid-Atlantic storm, a forecast of swell conditions on a European beach could be made some time in advance. There were various refinements to the method. For example, allowance could be made for the way that the waves might propagate, or travel, from the storm to the shore, including the effect of winds in the intervening space slowing them down or speeding them up. The same methods, now run by computer programmes, are used today to predict swell conditions that are posted on surfing websites.

To test the mid-1940s' predictions, routine wave measurements were made at various sites on the coast of the UK. Most of these were observations of wave height and period made by trained observers. These were supplemented in a few locations with measurements by instruments that could record the waves continuously. The first proper measurements of long swell waves travelling across oceans were made about this time, in the obvious place to study waves in this country: in Cornwall, near Land's End, where there is an uninterrupted run for waves travelling in from the Atlantic. The height of the water surface was measured with a pressure sensor placed in a metal frame laid on the seabed. As the waves passed over the sensor, they generated a varying electrical current, which was sent ashore along a submarine cable.

The people making these measurements (staff of the Admiralty Research Laboratory based at Teddington in London) realised that to test the forecasts they needed to be able to separate out waves with different periods, or wavelengths, from their records, since each wavelength would travel at a different speed and arrive at a different time. This was difficult because the swell waves were mixed up with other swell waves of slightly different period as well as with locally generated waves. The solution they came up with to solve this problem is breathtaking in its ingenuity, even after the passage of many years.

The trace of the pressure measurements (equivalent to the height of the water surface) wiggled along the centre of a long piece of chart paper. They arranged matters so that the paper was white on one side of the trace and black on the other. Any strip across the paper (that is, at a set time) was therefore divided into black and white parts in proportion to the height of the sea surface. A length of the chart paper, equivalent to about 20 minutes' worth of data, was then attached to the rim of a large wheel and a light was shone on to it such that it covered a small rectangle stretching across the paper. A photocell picked up the reflected light and created a current proportional to the amount of white paper, and so the height of the surface, at that time. When the wheel was set spinning, the output from the photocell was a repeated, high-speed version of the record of fluctuating pressure.

They connected the photocell to an instrument called a vibration galvanometer, which responded to an electrical signal with a particular period and ignored other periods. The mechanical equivalent of this device would be a pendulum. If you force a pendulum at its natural period, you generate a large movement; forcing of any other period will move the pendulum, but not very much. If the fluctuations of white and black on the photographic paper passed the photocell at just the right intervals of time, there would be a strong response from the galvanometer. Otherwise, the galvanometer needle would stay still. The clever thing was then to spin the wheel up to a fast speed (several revolutions per second) and leave it. Gradually, as friction slowed the wheel, the different periods in the pressure record came into tune with the galvanometer. The observers were able to plot out the strength of the signal corresponding to each wave period from the speed of the wheel and the deflection of the galvanometer.

This brilliant piece of practical analysis, before the availability of computers, was producing something called a *wave spectrum*: a plot of the height of waves against their period. A wave spectrum separates out the swell waves from the locally generated sea: the swell waves that have come from a distant storm have a long period and are clustered at one end of the spectrum. The spectrum also pinpoints the period of the swell. The observers noticed that the principal period of the swell waves arriving at their Cornish beach gradually changed with time: the period grew slightly less, on average, from one day to the next. They put this down to the fact that the different wavelengths were travelling at different speeds. The storm was creating waves with lots of different lengths and as these waves spread out from the storm, the long waves travelled fastest, outstripping their shorter-wavelength companions in the race to the shore. It was therefore the long waves that arrived at the beach first, followed in the next few days by shorter waves that had come from the same storm.

The scientists devised a simple way for working out how far from their Cornish observing site the storm must be. They made a plot (which they called a swell propagation diagram) with the distance from the shore on one axis and the time in days on the other. Waves with a particular period would have their own group velocity. Their rate of change of distance with time, the speed of wave approach, would equal the group velocity and so would produce a line with a known slope on this diagram. The time of arrival at Cornwall was known, so lines could be drawn on the diagram for each wave period. The place that these lines intersected must be the position of the storm.

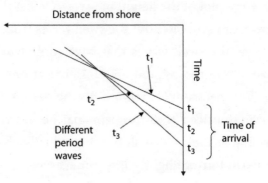

Swell propagation diagram for working out
the distance to a storm

They could check their weather charts for the weather system that must have generated these waves. Early on in this work it became apparent that some of the swell waves arriving at the Cornish observation site had travelled enormous distances. On one occasion, the waves appeared to have travelled a distance of 7,200 sea miles (one-third of the way around the world) and the scientists wondered if this could possibly be correct. There is a great circle route running out from Cornwall into the ocean for this distance, squeezing between Africa and South America and ending up in the South Atlantic near the island of South Georgia (see picture on page 112). It was possible that the waves could have come from a storm there, although it seemed remarkable that the waves could have travelled so far and still be detected. We now know that ocean swell waves can travel for thousands of miles across oceans, losing very little energy on the journey until they come crashing ashore.

There was one puzzling detail in the wave records. The period of the waves travelling from a distant storm grew less from day to day on *average* but it also fluctuated up and down in cycles of about 12 hours. The scientists thought that these fluctuations were produced by something that happened to the waves right at the end of their journey, when they reached the continental shelf. As the waves travelled from the ocean on to the shelf, they encountered significant tidal currents for the first time. A current opposing the waves would slow them down; the waves following behind would catch up and the period of the waves would be shortened. Conversely, if the

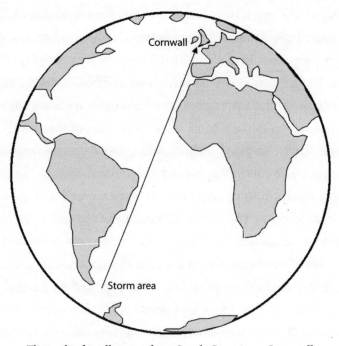

The path of swell waves from South Georgia to Cornwall.

tidal current was flowing with the waves, it would speed them up and lengthen the period. The tides would impose a half-daily rhythm on the period of swell waves seen at the shore. This effect wasn't really important to the wave forecasts but the fact that they could see it at all gave the observers confidence that they could detect small changes in the nature of the waves arriving at the beach.

The wave records were analysed by three young scientists working for the Admiralty Research Laboratory, NF Barber, J Darbyshire and F Ursell. George Deacon, the leader of the group and the first director of the National Institute of Oceanography, wrote about their work a few years later:

> *The observations and wave-recordings were examined on the spot by Barber, Ursell and Darbyshire, whose names will go down in the history of the subject. Mathematicians and observers had come together ... and reached the conclusion that no fundamental advance could be made in the study of waves until a method was developed by which a complex wave record could be resolved into its individual waves.*

Faith and I visited Pendeen to see the place where the work was done 'on the spot'. It's a remote place with a lighthouse, a few houses and the Cornish coast path winding away as far as the eye can see towards the old copper mines further along the coast. According to Deacon, the measurements were made 1,200 metres (4,000 feet) to the north-west of this lighthouse in 31m (17 fathoms) of water. There's no reminder now, though, of the contribution to oceanography and the war effort that was made here. Any huts and equipment are long gone. The lighthouse would have been familiar to the staff from the Admiralty Research Establishment who were based here, though. It was commissioned in September 1900 and stands four-square on the cliff, looking down on what was, on the day we visited, a spume-covered and confused sea whipped up by a strong south-westerly wind.

Pendeen lighthouse.

On the face of it, this north coast of Cornwall was a slightly odd choice for the measurements. Further round the headland, towards Land's End, would have had a more straightforward view out into the Atlantic and faced directly towards the swell coming from the ocean. Maybe there were logistic (or security) reasons for the choice of Pendeen (later, the measurements moved to Perranporth, even further east). More likely, it didn't matter: the Atlantic swell would have curved around the shallow waters surrounding Land's End and headed towards the north coast of Cornwall. As these long waves reach the end of their journey and move into shallow water, their speed depends more on the depth of water than on the length between wave crests. The part of the wave in shallow water would have travelled more slowly than the part in deeper water and the crests would have swivelled round to head directly towards shore. This refraction of swell in shallow water is the reason we normally see these beautiful regular breakers approaching directly towards the shore.

As with so many things, the demands of war had provoked great advances in knowledge. We knew much more about the ways that waves behaved in 1945 than in 1939. For several decades afterwards, research on ocean waves remained at the core of the science of physical oceanography. Effort was applied to improving measurement techniques and finding ways to analyse the data so that a record of sea level heights could be converted into easily understood values of a mean height and period. The concept of 'significant wave height' came to be widely used. This is the average height of the highest one-third of the waves and is close to what is reported by trained observers looking at the sea (our eyes tend to pick out the largest waves and ignore the small ones). Work on the theory of waves also focused on trying to find a formula to fit the spectra of waves that were being measured and to use this to help with wave prediction. Gradually, though, all but the hardest problems to do with ocean waves were solved, leaving just the toughest nuts about wave generation to crack. Attention drifted away to easier and more pressing problems in other areas of oceanography.

There is a photograph on the internet of the wartime wave research group at the Admiralty Research Laboratory. The photograph shows 13 smartly dressed people, 7 standing at the back and 6 sitting in the front. I think it must have been taken after the team had begun to split up: Barber and Ursell are absent. NF Barber, I know, moved to New Zealand in 1950 and wrote a brilliant book about the physics of waves on water. Even with these absentees, the photograph is remarkable because it shows, gathered together in a small group on the front row, the people who were to shape oceanography in Britain for the next generation.

At the time there was no research institute dedicated to oceanography in the country and only one university department, at Liverpool, but things were about to change. George Deacon, head of the group, sits fourth from left on the front row. Next to him, on his right, sits Henry Charnock, who was to take up the new chair of physical oceanography at Southampton University. On Deacon's left sits Ken Bowden, who would become head of oceanography at Liverpool. Jack Darbyshire sits next to Bowden at the end of the row; he would be the first professor of physical oceanography at Bangor in North Wales. Jack was a native of Blaenau Ffestiniog and when he turned up for interview, he delighted his interviewers by answering their questions in Welsh.

George Deacon himself was appointed director of the new National Institute of Oceanography, where he built up a world-class research institute by repeating the trick of attracting bright people and giving them the conditions in which to flourish. I met Sir George Deacon (he was knighted in 1977) after he had retired and as I was just beginning my career. It was only a brief meeting, but I keep the impression of a wise and kindly man who was genuinely interested in what a young scamp of an oceanographer was planning to do.

There is one aspect of water waves that we haven't mentioned and which, for many people, is the most important and depressing thing about them. They make us sick. Seasickness is a dreadful affliction: people sick at sea often don't care if they live or die. There are, thankfully, medicines that are effective in most (but not all) cases. The problem is that people are overconfident about their ability to cope with *mal de mer* and leave it too late to take the pills.

CHAPTER 5

A NIGHT AT 'THE ALMA'

ONE OF THE MOST TRAUMATIC EVENTS in the relationship between these islands and their neighbour, the sea, occurred in the winter of 1953. Towards the end of January in that year, strong and persistent north winds blew water into the North Sea faster than it could escape and the sea level steadily rose. On the night of 31 January, the high sea level, coupled with a large tide and storm waves, overwhelmed the coastal flood defences of south-eastern England, Belgium and Holland. Sea water surged inland with a force strong enough to carry away buildings and knock trains off their tracks. Most of the east coast of England and Scotland was affected, but the damage was greatest in the south-east – Essex in particular, where the sea level was highest. To compound matters, many houses on the Essex coast at this time were flimsy affairs, intended only as summer homes.

The flood was the worst peacetime disaster in 20th-century Britain. Over 300 people lost their lives in this country and more than 2,000 people were killed in Belgium and the Netherlands. Flood defences were breached in over 1,000 places and 24,000 homes were damaged, of which 500 were totally destroyed. There was extensive damage to farmland, with tens of thousands of animals killed, and 200 industrial sites were inundated. The total cost of the damage has been estimated as £1.2 billion at today's prices.

Although living memories of that night are fading, there are a number of visible reminders in the towns and villages of south-east England. The Alma Inn stands in the old part of Harwich, a mile or so from the ferry port where travellers leave for the Hook of Holland. The Alma is a cosy and welcoming pub, originally a merchant's home. It's the sort of pub-cum-hotel where you pick up your key from the bar. Then, if – like me – you are thirsty after a long journey and your attention is caught by the line of hand-pumps on the bar, you can take a pint over to find a quiet corner in which to sit for a while. A small plaque fixed to the wall by my table caught my eye. It looked like an official fixture, bearing the name Borough of Harwich, and it carried a mark showing the level that the 1953 floods had reached inside the pub. The mark is about 1 metre (3 feet) above the bar floor.

The Alma Inn in Harwich as it looked in the 1960s.

Sitting that evening in the Alma, watching customers at the bar ordering food and drinks, it was hard to imagine the events of that night in 1953 as sea water swirled around inside the pub. The flood reached its peak in Harwich at about midnight, but customers leaving the Alma at closing time (10pm in 1953) on Saturday night, 31 January, would have noticed that something was amiss. As they drained their glasses and stepped outside the warm pub, they would have had to brace themselves against the strong and very cold north wind. The sea in the harbour was already rough and boats were banging against each other. By 10.30pm, water had reached the top of the harbour wall and was flowing over the quay, even though high tide was not due for another two hours. By midnight, the sea was surging inland and much of Harwich was underwater.

Eyewitness accounts of the flood can be found in the Essex county archives and newspapers of the time; many are available online. Ruby Cooper-Keeble, 12 years old at the time of the flood, was in bed when she was woken by 'the water rushing up the stairs with enormous force. The lights went out and it was very dark. My mother opened the window. The water was halfway up the house and still rising.' Ruby, along with many others, spent the night upstairs watching the water flow along the street below. Many did not survive, drowning in the floodwaters or dying from exposure. It was a bitterly cold night and, in many homes, coats and warm clothes had been soaked by the floodwaters. Relief came, gradually, the next day as rescue boats arrived to ferry people to community centres where tea and hot food were available.

Thankfully, the meteorological conditions that led to the floods of 1953 are rare, but they were not, at the time, unprecedented. Twenty-five years earlier, in January 1928, there was another serious flood affecting the Thames Estuary. A common feature of both these events was gale-force northerly winds produced by a low-pressure system travelling over the north of Scotland. In both years, the low pressure travelled southwards into the North Sea (instead of following its usual route across the top of Scotland and eastwards to Scandinavia). This meant that the duration of the strong north wind was extended and also that gales were brought to bear directly on the shallow waters of the southern part of the North Sea.

A raised sea level at the coast produced by extreme weather conditions is called a *storm surge*. The 1953 storm surge was particularly devastating because it arrived at the coast of south-east

England at the same time as high tide and the coastal defences couldn't cope with the surge on top of the high water. Few parts of the Irish and British Isles are completely immune to storm surges, but the most vulnerable places are low-lying land facing a large stretch of shallow sea over which storm winds can act. One of the most devastating floods that we know about happened in the upper reaches of the Severn Estuary where, in 1607, between 500 and 2,000 people lost their lives in small villages and remote farms on the Somerset and Monmouthshire levels either side of the Severn. There is some question about what, exactly, caused that flood; it has been suggested that it was a tsunami, but a storm surge seems more likely. The level reached by the 1607 flood is recorded on the walls of local churches (there seems to be an abiding interest in commemorating extreme water levels). Chiselling the words 'flood level' into a church wall would have required a steady hand in the knowledge that a spelling mistake would be difficult to correct.

Following the North Sea floods of 1953, action was taken to protect the country from the worst effects of future storm surges. Physical improvements, such as higher sea walls, could be made almost immediately. Some years later, but directly as a result of the events of 1953, the Thames Barrier was built to protect London from flooding by storm surge. Better observations of sea level were also required, so that surges could be tracked as they moved around the country (a national network of tide gauges was created to do this job). Work also began on an effective forecasting system to provide warnings of future surges and allow time to evacuate people from threatened areas.

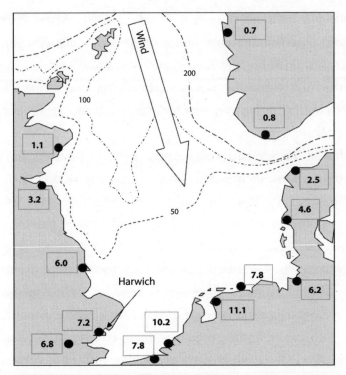

*The North Sea during the storm surge of 1953. Contours show
depths in metres; figures in boxes are the maximum height
the surge reached in feet.*

The map shows the level that the surge reached in 1953, taken
from a report prepared by JR Rossiter, the director of the Liverpool
Observatory and Tidal Institute, soon after the flooding. Surge
levels were calculated as the height of the sea above the level on a
calm day. The greatest surge levels, of around 3.5 metres (11 feet),
were observed in the southern part of the North Sea where the water
is shallowest and where the floodwaters are bottled up. Rossiter's
report was thorough and produced astonishingly quickly – it was

available within months of the event. It provided the basis for attempts to understand what had happened and to build capacity for predicting and responding to future storm surges.

History lies in many layers on these islands. In the 16th century, the building that currently houses the Alma Inn was a merchant's house, home to the Twitt family. In 1593, Sara Twitt married a young man, Christopher Jones, who lived across the road. Jones was a sea captain and, soon after his marriage, he sailed with his ship, the *Mayflower*, taking the Pilgrim Fathers to America. In the years that followed, Harwich grew in importance as a naval base and a centre for trade with Europe, particularly the Netherlands. Sara's family home passed down the generations and in about 1859 became the Alma Inn, taking its name from the Crimean war battle fought a few years earlier. The British Army was badly prepared for that war. Florence Nightingale used innovative graphical techniques, including pie charts, to convince the army authorities that they were losing many more soldiers to disease and lack of basic medical care than to enemy bullets.

There are two main causes of storm surges. The first is a dramatic fall in atmospheric pressure. As the barometer falls, sea level rises – a process known as the inverse barometer effect. To see how this works, imagine first an ocean covered by a blanket of atmosphere in which the pressure is everywhere the same. The weight of the atmosphere presses down uniformly on the ocean and the sea surface will be flat. The pressure at any given depth below the surface is the sum of the air and water pressure, and this is the same everywhere.

Now allow atmospheric pressure to change from place to place, as it does in the real world. Where air pressure is high, the combined air and water pressure below the sea surface will also be high and water will flow from this point towards places where the atmospheric pressure is low. This flow of water makes the sea level fall at places where the atmospheric pressure is high and rise where the atmospheric pressure is low, creating the inverse barometer effect. The effect is exactly the same as you see when you sit on an airbed: the bed sinks where you squash it and rises everywhere else. The flow of water from places of high to low atmospheric pressure will continue until the sea surface has adjusted enough to cancel out the variations in atmospheric pressure. Because sea water is much denser than air (it is about 1,000 times denser), a relatively small change in sea level is sufficient to offset the effect of the changes that occur in the air pressure. If the ocean has plenty of time to adjust to differences in atmospheric pressure, a fall in air pressure of 1mbar will create a rise in sea level of 1cm.

During the 1953 storm, there was a low-pressure system in the southern part of the North Sea and this would have contributed to the rise in sea level that produced the storm surge. The inverse barometer effect was not the main cause of the surge, however. Mostly, the damage was done by the strong northerly winds driving water into the North Sea.

When the wind blows over the sea, it exerts a force that drives the water along in the wind direction. In the open sea, the effect of Earth's rotation makes the current veer off at an angle to the wind, but near the coast the current at the sea surface will usually go directly with the wind. If the wind is blowing into an enclosed bay, the water level builds up in the bay until the extra pressure created by the raised water level balances the force of the wind. We can estimate the height of the surge by balancing the wind force against the pressure head created by the piled-up water. The idea of looking for a force balance, by matching the two most important forces acting on the sea in a given situation and seeing what we can learn from that, is a useful one in oceanography. When forces come into balance, the net force becomes zero and (according to Newton's second law of motion) accelerations stop: the situation settles down into equilibrium. The calculation of the force balance in a storm surge is not a difficult one; it is the sort that can be carried out on a piece of paper the size of (since I am sitting here in the Alma) a beer mat.

First, we need to know the force that the wind exerts on the sea. It's not immediately obvious what this might be. How can a horizontal wind exert a force on a similarly horizontal water surface? A wind blowing over a glassy calm sea, for example, could be expected to slide over the water without exerting any force at all on the sea. But even in this situation, small variations in air pressure created by atmospheric turbulence will produce unevenness in the sea surface. The wind can then push on the little wavelets that project upwards into the air. As it does so, the wind exerts a force and transfers some of its momentum into the sea. The water speeds up and the atmosphere slows down (or replaces the lost momentum from higher levels of air). As the waves grow, this momentum transfer mechanism becomes more effective. The rate of transfer of momentum, expressed per unit area of sea surface, is called the *wind stress*. Wind stress has units of force per unit area, the same as pressure, but the force of the wind acts tangentially to the surface of the sea and drags it along rather than presses against it. The wind itself drags along only the surface layer of the sea, but this then drags the layer of water below that, and so on, so that the horizontal force of the wind is ultimately passed downwards towards the seabed.

Observations tell us that wind stress at the surface increases with the square of the wind speed. We can see how this might be so by considering the way that momentum is transferred from the air to the water. We can divide up the wind, in our imagination, into small packets of air, each one – say – the weight of a tennis ball. These packets hit the face of a wave on the sea surface and bounce off or come to rest, transferring their momentum to the water as they do so. The momentum in the air packets is the product of their mass and velocity, and the number of tennis-ball packets that hit the water in a set time (say a minute) is also proportional to the wind velocity. The rate at which momentum is transferred from the air to the sea therefore increases as the *square* of the wind velocity. Because the density of air is much less than that of water, the speed at which the water moves as it picks up atmospheric momentum is much less than the wind speed. The surface layer in a wind-driven current moves at a speed that is just a few per cent of the wind speed.

You might get the impression from the last paragraph that the rule about the force of the wind on the sea being proportional to the square of the wind speed is imprecise, and you'd be right. It's not a law of physics. Instead, it is a rule that is based on observations: we call it an *empirical* law. The constant of proportionality between the wind force and the square of the wind speed sometimes has to be adjusted to fit observations. That tells you that we don't really understand what's going on. Nevertheless, the rule usually works pretty well and is certainly good enough for the calculation we are going to do.

If we measure the wind speed in m/s, the force of the wind, in newtons, acting on one square metre of sea surface is about 10^{-3} times the square of the wind speed. Wind stress can only act directly on the sea surface, but the momentum is then transferred downwards into the water below. Oceanographers like to deal with the force acting on unit mass – 1kg – of water. To get the force per unit mass we have to divide the surface stress by the number of kilograms of water below the surface. The wind force, in newtons per kilogram, works out as about 10^{-6} times the square of the wind speed and divided by the water depth in metres. Observations at light vessels in the North Sea at the time of the

1953 floods tell us that the peak wind speed was about 30m/s. The depth of the North Sea varies from about 20m in the south to over 100m in the north; we can take an average depth of 40m. The average force of the wind driving water into the North Sea at the time of the 1953 floods is then 10^{-6} times 30^2 divided by 40, or 2.3×10^{-5} newtons per kilogram. It's a very small force, but it is similar to other horizontal forces in the sea and so is important. We can also note that the wind force will be somewhat less in the northern North Sea, where the water is deeper, and greater in the shallow southern North Sea.

The second thing we need to know is the horizontal force created by the head of water raised at the coast – the force that ultimately balances the effect of the wind. The pressure at a depth in the sea depends on the weight of water above that depth (plus the atmospheric pressure acting on the surface). As we go down into the sea, the pressure increases. Divers experience this first-hand and we can lower instruments to tell us how the pressure in the sea varies with depth (these commonly use a piezoelectric crystal, which generates a small electric current when squeezed). Marine animals are able to cope with the pressure change with depth: whales, for example, have a flexible body structure that can absorb the increase in pressure as they dive.

A curious thing about pressure in water is that, although it is created by Earth's gravity acting exactly vertically down, the pressure at any point in the sea is *the same in all directions*. This follows from the fact that liquids cannot resist being squeezed. If we take the cap off a toothpaste tube and press the sides of the tube, toothpaste squirts out of the nozzle. The toothpaste follows the path of least resistance and moves in the direction where the pressure is smallest. If a parcel of liquid, of any shape, is *not* squeezed equally in all directions, it will elongate in the direction of least pressure. The fact that water in a glass can stay still without pieces stretching out in random directions tells us that the pressure must be acting equally in all directions. A small parcel of water in the glass is squeezed equally on all sides and it also presses equally in all directions on the water (or glass) surrounding it.

It has to be admitted that an explanation along the lines of 'this is what we see and so that must be so' is fundamentally unsatisfactory. The ideas of the last paragraph don't explain *why* the pressure in the sea acts equally in all directions. I think the answer to that question has to do with the way molecules in a fluid interact with each other: they can press equally well on all surrounding molecules and also slide over other molecules if the pressure is not the same in all directions. But that's more than we need to go into in this book. The important thing for us is that the pressure in the sea increases with depth below the surface and acts equally in all directions. The vertical part of the pressure presses against the seabed, or holds the column of water up, but the horizontal component of the pressure force is available to balance other horizontal forces, such as the force of the wind.

When the water surface is flat and the water is of uniform density, the horizontal part of the pressure is the same on any horizontal level below the sea surface. If the water surface is sloping, however, the pressure is greater under the high part of the slope and there is a horizontal variation in pressure, called a pressure gradient force. The pressure gradient force acting on unit mass of water is equal to the slope of the water surface multiplied by the acceleration due to gravity and, because it is produced by a surface slope and so just depends on the difference in sea level between one place and another, it is the *same at all depths*.

We now have the tools we need to calculate how high the surge will get for a given wind speed. The average force exerted on the water by the wind at the peak of the 1953 surge was 2.3×10^{-5} N/kg. If this force is balanced by a pressure gradient force (also in N/kg) equal to the surface slope multiplied by the acceleration of gravity (and we take the acceleration of gravity to be 10m/s^2, which is close enough for a beer mat calculation), then the *average* slope of the surface must be about 2.3×10^{-6}, or 0.23cm per km. The length of the North Sea from the north of Scotland to the Thames is about 1,000km. A slope of 0.23cm per km acting over a horizontal distance of 1,000km will raise the level of the sea surface on the weather shore by 2.3 metres. This is our estimate of the height of the surge. The maximum observed level of the surge on the Dutch coast during the 1953 surge was somewhat higher than this, 11 feet or about 3.4 metres.

This calculation tells us something fundamental about the sort of shores that are vulnerable to storm surges. First, the strength of the wind force increases with the square of the wind speed and inversely with the depth of water. Wind force is most effective, for a given wind speed, over shallow water. The southern North

Sea is particularly shallow, not much more than 20 or 30 metres (66–98 feet) deep on average and it is here that the gradient of the sea surface is at its steepest to match the wind force. Second, the height of the surge is equal to the slope of the surface multiplied by the length of the basin over which the wind acts. Storm surges will be particularly problematical on coasts that face a prevailing wind blowing over a long stretch of shallow water. The North Sea during a northerly gale fits this description very well.

You might be wondering how the currents in a surge vary with depth. Does the water just below the sea surface move in the same way as the water just above the seabed? The answer is no. When the wind force has come into equilibrium with the pressure gradient, the net flow of water (averaged from surface to bottom) is zero, but the water is moving. There is a current directed onshore at the surface and one in the opposite direction, offshore, along the bottom. The wind drives a surface layer of the sea in the direction of the wind. When this moving water reaches the coast, it has only one place it can go. It dives down and returns in the offshore direction as a flow along the bottom travelling towards the open sea. The pressure gradient created by the surface slope now has the extra task of overcoming the effect of bottom friction on the offshore flow and, as a result, the surface slope increases. A bit of extra calculation shows that the need to drive the bottom current against the effect of bed friction steepens the surface slope by 50 per cent. This increases our estimate of the size of the surge from 2.3 metres to about 3.4 metres, or 11.2 feet, now very close to what was seen on the southern shore of the North Sea in 1953.

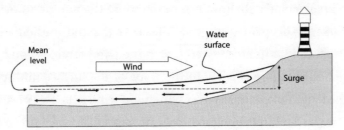

Cross-section of an idealised storm surge. The slope of the sea surface depends on the depth of water and is greatest in shallow water. The arrows beneath the water surface show how the water circulates.

We could refine our calculation by allowing for variations in the depth of the North Sea but there is really not much point. The thing is that any calculation we can perform with a pencil and beer mat will have to assume that the surge has reached equilibrium, with the force of the wind balanced by the pressure gradient of the slope. In the surge of 1953 (and any real surge), the wind constantly changed direction and speed and the water surface had to adjust to these changes. Surges in the North Sea are created by intense low-pressure systems passing north of Scotland. Before the low pressure arrives, the wind is from the south and water is actually sucked out of the North Sea, lowering the sea level on the south-east coast of England. A lowering of sea level before the 1953 surge was apparent in the contemporary tide gauge records. Then, when the wind direction shifts to the north and starts to blow strongly, water is driven into the North Sea and travels as a single wave crest – a surge – down the eastern coast of Great Britain, taking about 12 hours to cover the length of the coastline from the north of Scotland to the south of England.

A great volume of water – about 400 cubic km – was forced into the North Sea from the Atlantic at the height of the surge. Some of this extra water would have escaped through the Dover Strait into the English Channel. The gap between England and France here is a narrow one, though. Calculations made at the time suggested that water escaped in this way at a rate of about 5 cubic km per hour. The Strait of Dover is not a great safety valve and wouldn't have had much of an effect on the surge in the North Sea. Most of the water travelling southwards in the surge was reflected off the southern coasts of the North Sea – the coasts of Belgium and Holland – and then the surge travelled northwards, back out of the North Sea, losing energy and height. As a result, some places experienced two surges, one as the wave passed coming into the North Sea and a second, weaker one, as it passed on its way back out again.

Catching the full dynamics of a storm surge as it responds to changes in force and bounces off coasts is more than we can do on the back of a beer mat. It was more than anyone could do in 1953 but, gradually, theoretical models, combined with observations of sea level, were developed to forecast future surges. In the 1950s, the first computer models designed to forecast the weather were becoming available. For predicting a storm surge, these weather-forecast models were combined with sea forecasts that predicted the response of the water to the force exerted by the wind.

In 2017, Kazuo Ishiguro became the eleventh British winner of the Nobel Prize for literature, joining a list that includes TS Eliot, Winston Churchill and Rudyard Kipling (for a small nation, Britain has an exceptional number of Nobel laureates: across all categories, the tally is second only to the United States). Ishiguro's award is linked to the story of this chapter by the reason he came to be in England. He was born in Nagasaki, Japan in 1954. His father, Shizuo Ishiguro, was a pioneer in using computers to simulate the flow of water. The director of Britain's National Institute of Oceanography, George Deacon, met Ishiguro senior in Japan and invited him to England to work on the problem of predicting storm surges. Writing this makes me think how times have changed in the field of staffing science establishments. Deacon seems to have had a free hand to recruit the talent that the country needed; there was no need to go through the laborious selection procedures that would surely get in the way now.

Anyway, Ishiguro accepted the invitation and moved with his family, future Nobel-laureate son and all, to Surrey in the late 1950s. The computers he built were not the same sort that we are familiar with today. They were *analogue* computers; the flow of water in the sea was represented, in the computer, by an electrical current in a wire. Changes in the voltage, or potential differences, along the wire were equivalent to changes in sea level. Friction on the flow could be simulated by introducing electrical resistors into the wiring and Ishiguro found ways to drive the electrical currents with forces that represented the effect of the wind and of Earth's rotation.

The machine could allow for variations in water depth in the North Sea and changes in wind speed and direction; the predicted water levels and their changes over time were displayed on oscilloscopes. This was an attractive approach in the days when digital computers were still in their infancy. Shizuo's machine could run a simulation of the development of a storm surge in the North Sea in a matter of seconds – much quicker than a digital computer of the time. When Ishiguro retired, he continued to develop his computer at home in his small garden shed. This important early solution to the problem of predicting storm surges is now on display in the Winton Gallery in the Science Museum in London.

The future for computers, though, was digital. Ishiguro's machine was replaced by numerical solutions to a combined weather and sea model of the North Atlantic and European shelf seas. The sea is divided, in the computer model, into a series of rectangular boxes and the motion induced by the wind force acting on each box is calculated. Two equations are used in these models. The first is the equation of motion, Newton's second law, which allows the water current to be calculated from the applied forces. The second is the equation of continuity, which keeps track of the volume of water in each box from reckoning the flows in and out of the box. The computer simulations can allow for the effects of the Earth's rotation and bottom friction on the flow.

As weather forecasts and our understanding of ocean dynamics (including the thorny problem of momentum transfer from the atmosphere to the ocean) have improved, storm surge modelling has become a reliable tool. We no longer have to wait upon events; a serious surge can be predicted several days in advance. Warnings can be issued and, if deemed necessary, the Thames barrage can be closed. For a while, a friend of mine – Kevin Horsburgh – was in charge of storm surge modelling at the National Oceanography Centre in Liverpool. I liked to think he had a button on his desk that closed the Thames barrage when his computer bubbled over, but he told me it wasn't that simple. The important thing is that the combination of reliable computer forecasts, improved flood defences and mitigation procedures has been a success. In 2013 there was a storm surge in the North Sea of similar magnitude to that of 1953: 18,000 people were evacuated and nearly 3,000 properties flooded but there were no deaths and the cost of the damage was much less than in 1953.

There are many subtle interactions between the atmosphere and the surface layers of the sea but there is room in this chapter to tell of just one more. You may have noticed that when a strong wind blows over an open stretch of water, it can produce parallel lines of foam, called *windrows*, lined up with the wind direction. The first person to take a scientific interest in windrows was Irving Langmuir, who spotted them while he was leaning on the rail of a transatlantic liner in August 1927. I can do no better than to quote his own words on the subject:

> *when about 600 miles from New York on an Atlantic passage to England I noticed that there were large quantities of floating seaweed, most of which were arranged in parallel lines with somewhat irregular spacing ranging from 100 to 200 metres. These lines, parallel to the wind direction, which I shall call streaks, often had lengths as great as 500m.*

On the following day, the wind changed direction by 90 degrees and Langmuir noticed that the windrows responded quickly to this, soon lining themselves up with the new wind direction 'although the waves continued to move in the old direction'.

I find these observations delightful because they were made, not by someone who had a professional interest in this but by someone who was observant, curious and had the intelligence to realise there was something to explain here. Studying the ocean can be frustrating. Because you are seeing nature in the raw, it is not possible to separate out what you want to see from all the other things going on, as can be done in a well-designed laboratory experiment or in a computer simulation. Sometimes, measurements have to be made laboriously and for a long time if you want to see how things change with the seasons or longer periods. Results in oceanography are often ground out slowly and oceanographers are left behind, in publication rates, by colleagues who can complete a laboratory experiment in a day. It is only right that there should be compensations for these difficulties. The big compensation, for me, is that nature is continually providing a show for you to interpret. New things can be discovered on a sea crossing or on a stroll along the beach. And that opportunity is open to everybody.

Even though it had nothing to do with his job at the General Electric Company in New York State, Langmuir continued to take an active interest in windrows. He noticed that they formed on a lake near his home and observed them closely, using natural and artificial floats (light bulbs were a favourite) to see how the water was moving. In the autumn, he noticed that falling leaves settling on the surface of the lake moved towards the foam lines, where they collected; some of the leaves then moved vertically down, disappearing from view. This observation confirmed his opinion that the streaks were formed by surface currents converging from the sides; the water then sinks downwards, leaving floating material to collect on the surface. The sinking water spreads out at depth and returns to the surface along lines midway between the foam lines, where it then moves along the surface back to the streaks. This circulating motion is superimposed on the general movement of the surface water driven by the wind, so that a parcel of water has a corkscrew motion as it both circulates and travels along the windrows.

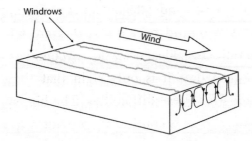

Circulation cells between windrows on the sea surface.

Langmuir also found that the size of the circulation cells (and so the separation of the windrows) was set, in summer, by the depth of the warm surface layer on the lake: the circulation went down as far as the bottom of this layer and no deeper. In fact, he concluded, the circulation was important in mixing this surface layer. In a particularly charming experiment, he laid a piece of string, buoyed by corks, across the windrows. He found that the string moved along with the wind quicker in the streaks and slower in the spaces between: the string took on a wavy shape. His interpretation of this observation was that the water upwelling between the streaks had received no previous momentum from the wind and so slowed down the flow when it reached the surface.

The remarkable thing about Langmuir's insights and measurements, made nearly 100 years ago with light bulbs and string, is that we have added precious little since to the pot of knowledge on the subject. The outstanding question now is what drives the circulation (now named after Langmuir) between the windrows. Current theory is that it is an interaction between lateral variations in wave height and the Stokes drift in the waves, but I find it difficult to see how this fits with Langmuir's noticing (all those years ago) that the windrows could follow a change in wind direction much more quickly than the waves. Nature guards its secrets jealously and it is taking particular care of this one.

CHAPTER 6

A HIGH TIDE AT CLEVEDON

The greatest rise and fall of the tide in these islands is seen in the upper reaches of the Bristol Channel where the counties of Gwent, Gloucestershire and Somerset converge. At Clevedon in Somerset, the tidal range (the vertical distance between low and high water) on a day of big tides is 14 metres (47 feet). That's higher than most houses in this country, including their roofs and chimneys. These are the largest tides in Europe and the largest in the world outside a few locations in North America. The big tides stamp their characteristics, in full measure, on this part of the Bristol Channel. The water is khaki-coloured and opaque: it is full of tiny mud particles stirred up by tidal currents. Low water exposes brown expanses of sticky gloop, the sort that clings to wellington boots and leaves their wearers floundering. Sometimes it is hard to say where the mud banks end and the muddy brown waters begin.

Clevedon is a handsome, prosperous town with large Victorian villas clustered in the streets around the seafront, a legacy of the days when merchants from Bristol chose to live here. The nurse Edith Cavell grew up in the town and a few doors down from her childhood home, a blue plaque announces that penicillin was developed there. The town has a beautifully restored Victorian pier; the only Grade I listed pier in the country. Inside the pier museum there is a model in which you can wind a boat up and down to follow the motion of the tide at the end of the pier.

The tidal range at Clevedon is undeniably large, but I have to confess to being slightly underwhelmed by the overall effect. Compared to Blackpool, where I have childhood memories of the sea seeming to go out for miles, the low tide at Clevedon doesn't retreat all that far down the shore. The reason for this difference may be attributed to my young, short legs on the Blackpool visit, but it is also to do with the slope of the beach. With a gentle beach gradient of 1 in 1,000, for example, a fall in tide of 10 metres will make the edge of the sea go out by 10km. That's the sort of thing that happens in Morecambe Bay, north of Blackpool, where great expanses of mud and sand are exposed at low tide. At Clevedon, the gradient of the shore is steeper. On the day I went with my daughter, Anwen, the high tide reached to the top of the beach but the low tide didn't quite reach the end of the pier, 310 metres out to sea. For a tidal range of 10 metres, that suggests that the beach slope here is more like 1 in 30.

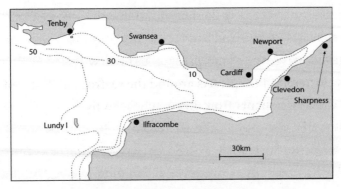

The Bristol Channel. Depth contours are in metres.

What is undeniably impressive, though, is the sheer energy of the tide in the Bristol Channel. The channel has the shape of a crumpled wizard's hat, with a width at the mouth near Lundy Island of 70km (43 miles) and a length, from Lundy to Sharpness, of 170km (106 miles). Each tide – that is, twice a day – up to 60 cubic km of water flow in and out of the channel, to make the tide rise and fall. Transferring this great volume of water in a limited time requires tremendous flows over a large area. The Bristol Channel experiences tidal flows of 3 knots (1.5m/s) or more: the waters outpace the Gulf Stream twice each tide, or four times a day, as they push the water in and out of the channel. It is these energetic flows that stir up the mud from the estuary bed and give the water its characteristic colour and opacity.

The information board in the pier museum explained that the large tides of Clevedon are caused by the tapering shape of the Bristol Channel. Anwen and I retreated to the Royal

Oak to consider this. A comfortable buzz of conversation and the clacking of dominoes accompanied us as we carried our pints over to a corner. The information board at the pier made sense up to a point. The tide is a wave moving over the water surface, generating currents that reach to the seabed. High water corresponds to the crest, or top of the wave, and low water to the trough. As the wave moves into a converging channel, the energy is compressed and the wave height, or tidal range in the case of a tide wave, will increase. The width of the Bristol Channel contracts by a factor of five between Lundy Island and Clevedon and we would expect that, for a wave travelling at constant speed without losing energy, the tidal range should increase by a factor equal to the square root of 5, or 2.24 (the square root comes into this because the energy in a wave depends on the square of its height). The actual increase in tidal range between Lundy and Clevedon is by a factor of 1.66, somewhat less than we would expect from the tapering of the channel. Moreover, the depth of the Bristol Channel also shoals inland from Lundy. The wave will slow down as it travels up the channel and this will also compress the energy as the parts of the wave behind catch up with the part leading the way. Including this depth effect, the increase in tidal range from Lundy to Clevedon should be by a factor of 3.1, nearly twice as much as is seen. The surprising thing, then, is not that the tide in the Bristol Channel is so great but that it is not actually much greater. I'd like to say that all this was clear to us as we sipped our beer in the Royal Oak, but we were enjoying the delights of the pub too much for that to be true. It did become clear soon after, though, when we could sit down with a map and do a few simple calculations.

The cause of the ocean tide is an ancient mystery that puzzled the early Chinese, Arabic, Roman and Greek civilisations. It was apparent to any careful observer that there must be a link between the tide and the movement of the moon. High tide at the coast occurs 50 minutes or so later each day, and so does moonrise and moonset. What's more, the tides are biggest twice each month when the moon is full or new. The moon obviously influences the tides in the ocean, but how can it do that, exactly?

The breakthrough came with the publication of Sir Isaac Newton's Law of Universal Gravitation in 1687. Newton realised that in order to make the solar system work in the way that it does, the sun, planets and moons must attract each other with a force that varies inversely with the square of their separation distance. Our own moon exerts a gravitational pull on the Earth and its oceans and the strength of this pull changes over the surface of the Earth, getting weaker with increasing distance from the moon. Once a month, the Earth and moon orbit each other about their common centre of gravity. The centripetal force required to hold the Earth in this orbit is provided by the moon's gravitational pull. The moon's gravity has exactly the right strength to do this job at the centre of the Earth but it is a little too weak in the Earth hemisphere facing away from the moon and a little too strong in the hemisphere facing the moon. These deficiencies and excesses (relative to the required centripetal force) in the moon's gravitational pull at the surface of the Earth are called *tidal forces*. Under the action of tidal forces, loose objects on the surface of the Earth will tend to slide towards the moon in one hemisphere and away from the moon in the other. If the Earth were turning slowly and was covered in an ocean, the waters of the ocean would move into two humps, one directly under the moon and the other on the far side of the Earth to the moon. A point on the surface of the turning Earth would then experience two high waters in a day as it passed through these humps, one when the moon was directly overhead and the other when the moon was directly underfoot. Newton had solved one of the age-old problems of the tide: why there are two high waters each day when the moon passes overhead just once.

In the time it takes for the Earth to turn once, the moon moves a little way around its orbit and the point on our turning Earth has to turn a little further to catch up with this motion. There are two high tides in a *lunar day,* the time it takes for the moon to return to the same place in the sky. A lunar day is 24 hours and 50 minutes and so the time interval between high tides is half this figure, 12 hours and 25 minutes. The average time between high waters in the Bristol Channel, and indeed most of the world, is half a lunar day: the tides are said to be half-daily, or *semi-diurnal*. The sun also creates tides in the ocean and these have a period, or time interval between high waters, of 12 hours exactly. As we move through the month, the moon and sun tides move in and out of phase. When they are in phase, and the sun and moon make a straight line with the Earth, there is a large tidal range called *spring tides*. These large tides are seen every 14 or 15 days. In between days of spring tide, there are days when the sun and moon make a right angle with the Earth; their tidal forces now cancel to some extent (not completely, because the moon's tidal forces are greater than the sun's) and this makes for smaller, *neap* tides.

In reality, the Earth is not covered in an ocean that piles up into two lumps under the action of the moon's tidal forces. Instead, the tidal forces act on water that is held in ocean basins and partly surrounded by continents. The effect of the small, but regularly applied, tidal force is to create a rocking motion on the surface of the ocean. The period of the rocking is the same as the period of the tidal forcing – so half a lunar day (or 12 lunar hours) in the case of the moon's tide. It is high water on one side of the ocean when it is low water on the opposite side and vice versa. In the North Atlantic, for example, the tide is high in Europe when it is low in America. Six hours later, it is high water in America and low water in Europe. The tides in the deep ocean can be most easily measured on islands. They are not very large; the range of the tide on islands in the middle of the ocean is typically half a metre or so. The largest tides on our planet are observed on coasts that are separated from the ocean by a continental shelf and are produced by a process called *resonance*.

The effect of resonance on a container of fluid will be familiar to anyone who has tried to carry a bowl of soup from the stove to the table. The natural sloshing period of the soup in the bowl is close to the period of the swaying of a walking human body; it is very easy to lose control and to spill the soup out of the bowl if you are not careful. In fact, thinking about this regular mealtime hazard, I realise that, when it's my turn to carry the soup, I try shortening my steps in an attempt to avoid this particular resonance effect.

The tide in the ocean creates a resonant effect on a continental shelf that has the right dimensions. The rising oceanic tide pushes the crest of a tide wave on to the shelf. The wave crest then travels

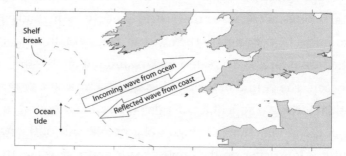

Tidal resonance will occur on a continental shelf when the time taken for a tide wave to cross the shelf to the coast and back matches the rise and fall of the ocean tide.

across the shelf until it encounters a coast. Here, it is reflected (coastlines seem to be very good at reflecting tide waves) and travels back out towards the shelf edge. If the crest takes exactly six lunar hours to make the journey from the edge of the shelf to the reflecting coast and back, it will arrive at the shelf edge at a time when the falling tide in the ocean is working to pull the crest off the shelf. The pull and push of the ocean tide then just matches the motion of the wave back and forth across the shelf, and the shelf comes into resonance with the ocean tide. Resonance will also occur on a shelf in which the journey time from the shelf edge to the coast and back is 18 or 30 lunar hours, in which case subsequent wave crests will arrive at the shelf edge at just the right time. On a resonant shelf, the ocean tide is continually feeding energy into the shelf tide and the tide waves travelling from the shelf break to the coast and back grow in size until they lose energy at an equal rate through friction with the seabed and imperfect reflection at the coast.

The north-west European continental shelf, as a whole, is close to resonance with the ocean's semi-diurnal tide and that is why we experience such big tides in these islands. In Chapter 4, we learned that the speed of storm-made waves crossing the ocean depends on the wavelength – the distance from one wave crest to the next. This rule does not apply, however, to tide waves. Tide waves have a very long wavelength – much greater than the depth of the ocean – and their speed just depends upon the water depth. They can travel fast. In the case of the approaches to the Bristol Channel shown in the picture, the speed of the tide wave crossing the shelf is about 140km/h (87mph). The distance from the edge of the shelf to the Bristol Channel entrance is about 400km (250 miles). The tide wave takes a little under six hours to make the return crossing from the ocean to the coast and back and we can expect a resonant effect on this part of the shelf. This is exactly what we see. The tidal range on an ordinary tide (that is, halfway between springs and neaps) increases from about 1 metre (3.3 feet) at the edge of the continental shelf to 5 metres (16 feet) at Lundy Island. This amplification of the tide by resonance on the shelf is an important part of the story of the big tides in the Bristol Channel: the tide has already been increased by a factor of five by the time it gets to the entrance to the channel. The journey time from the shelf edge to the coast and back becomes closer to six lunar hours (and the resonance becomes more exact), I think, if you include the trip into the Bristol Channel. But once in the channel, the wave behaves in a special way because of the shape of the coastline and the effects of friction.

The first person to think hard about the tides in the Bristol Channel was the British scientist Sir Geoffrey Ingram Taylor (1886–1975). GI Taylor was one of the great physicists of the 20th century and contributed his skills to the war effort in two world wars. He wrote elegant papers on the design of aeroplane propellers and parachutes and, to help with these, he learned how to fly and jump out of aeroplanes with a parachute on his back. He loved sailing and designed an anchor, for use by small boats and seaplanes, that dug into the seabed like a plough and was a big improvement on contemporary anchors, which were of the sort you can still see on some packets of butter. During the atomic bomb tests after the Second World War, Taylor worked out and published an estimate of the power of the bombs, which he had calculated from photos of the developing mushroom cloud published in a magazine. This confounded the authorities at the time, first because the information was classified and second because Taylor's answer was astonishingly accurate. (As an aside, the same principles that produce the mushroom cloud from an atomic bomb apply to the fate of dredged silt, which is sometimes taken out to sea to be dumped. The silt sinks quickly to start with and then spreads out laterally as it falls through the water, making an inverted mushroom shape.) In an appreciation written for the 100th anniversary of Taylor's birth, GK Batchelor wrote:

> *G.I. Taylor was a happy man who spent a long life doing what he wanted most to do and doing it supremely well.*

What a wonderful thing to write about anybody, and what a wonderful thing to have written about you.

Taylor published his ideas about the tides in the Bristol Channel in 1921, at the age of 35. He considered an idealised channel, tapering such that the width and depth both decreased steadily in proportion to distance from the mouth. He found that tides in the channel could be described by a particular kind of mathematical formula called a Bessel function. Bessel functions are used by engineers to describe the sort of waves that form in round containers: the circular waves that form on the surface of tea in a mug placed on top of a rumbling washing machine can be described by a Bessel function. Some engineers have circular testing tanks in which they can produce a large-scale version of this sort of wave. The wave is made by a sudden inward movement of the sides of the tank and, as the wave travels inwards, the energy is concentrated and the wave grows. A particularly energetic wave can squirt upwards like a fountain when it reaches the centre of the tank.

Taylor's calculations predicted that the tidal range in the channel would increase by a factor of 1.7 from Tenby to Clevedon, in good agreement with the observed increase in range. Somewhat surprisingly, the Bessel function solution doesn't depend on how quickly the sides of the channel converge. The solution for the converging channel is a portion of the full solution for a circular tank, like a wedge of apple pie cut from a full, circular pie. The pattern of the surface is the same for all portions, no matter how small or large. The wave amplitude increases towards the sharp point of the wedge in exactly the same way as it does in the full circle of the tank.

In Taylor's solution, the tides rise and fall in the Bristol Channel in unison. It is high tide everywhere at the same instant, and low tide everywhere six hours later. In the interval between low and high tide, water flows into the channel from the open sea. The water flows out again in the period between high and low tide. At both high and low tide, the current momentarily stops, or becomes slack, before reversing direction. The only variation in the tide along the channel, in Taylor's model, is that the tidal range gently increases from the mouth up to the head. This kind of tidal motion is called a standing wave.

This picture is very close to the tidal behaviour in the Bristol Channel, but it's not perfect. In the real tide, there is a time delay of over an hour between high tide at the entrance to the channel, at Lundy Island, and Clevedon (see picture below) and the current continues to flood into the channel for a short while after high tide. These variations from a standing wave are caused by the effect of friction on the tidal flows. Taylor did not include friction in his calculations, even though he was well aware of it and had already made some of the first calculations of the strength of frictional forces on tidal flows.

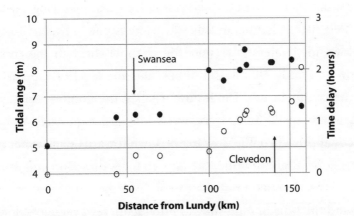

Tidal progression in the Bristol Channel. Black points show the mean range of the tide and white ones the delay in the time of high water after that at Lundy Island.

Taylor's solution can be modified to include friction but the result is a more complicated set of Bessel functions; it becomes harder to form a mental picture of what is actually happening. In my opinion, a satisfactory understanding of the way that nature works should be more than a set of mathematical expressions. There should also be an idea that a non-specialist can grasp and which, ideally, can be viewed as pictures rather than equations. The idea should be simple enough to hold the whole thing in your head and it should be memorable. It may not explain everything that is observed but, if it's right, it will account for the salient features. It gives you the opportunity to snap your fingers (if you can do that) and say, 'Ah, so that's how it works!'

I think there might be a number of ways of picturing the tide in the Bristol Channel in this way. Here is one of them. Starting at low tide, water flows into the channel through the mouth between Lundy Island and Tenby and the level of the surface in the channel rises. The inflow – called the flood – reaches its maximum speed halfway between low and high tide. As high tide is approached, the flow slows down towards slack water and then, soon after high tide, the flow changes direction and moves out of the channel, or starts to ebb. A force is needed, applied around the time of high tide, to provide these *changes* in velocity, or accelerations. The force has to slow down the flood current, turn it around and speed it on its way as it becomes the ebb. This force is provided by a slope (upwards, inland) on the water surface at high tide, as I've sketched on page 157. In a similar

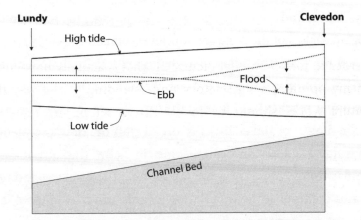

Profiles of the water surface along the Bristol Channel at four stages of the tide.

way, at low tide, the surface must slope down inland to provide the force required to turn the ebb into the subsequent flood. These slopes mean that the high tide at Clevedon must be higher, and the low tide lower, than that at Lundy and so the range of the tide must increase moving up the channel.

Exactly at low and high tide, the currents themselves are weak and frictional effects are small. But this is not the case when the flood and ebb currents are flowing at their fastest. During the maximum flood, a force is needed to drive the water against the effect of friction from the bottom and sides of the channel. The friction will be greatest in the upper reaches of the channel, where the water is shallow and the sides are close together. During the flood, the water surface slopes down in these upper reaches, making a pressure head to overcome the friction on the flow. As high water is approached, the currents slow down and friction dies away. Its job done, the slope in the upper reaches of the channel now levels off and the surface rises gracefully – like the flukes of a whale's tail – to catch up with the level at Lundy. But this takes time. By the time the water level in the upper reaches has reached the top of its trajectory, the level at the mouth of the channel is already falling. In consequence, high tide at Clevedon occurs some time after that at Lundy. A similar thing happens on the falling tide; the level at Clevedon chases the level at the mouth downwards and falls to meet it as low water is approached.

This picture is a good one, I think. It includes the important features in the shape and dynamics of the tide in the Bristol Channel and it explains the amplification and delay in the tide at Clevedon compared to the mouth. Also, there is another way of interpreting this picture that I quite like and which gets closer to the spirit of the information on the board at Clevedon pier. The shape of the water surface at any time can be regarded as different parts of a *wave* travelling into the Bristol Channel. The wave is not like any other that we have come across in this book: its wavelength and speed both depend on friction. Once we've grasped that idea, we can work out the important features of this wave. Only a small part of the wave fits into the space between Lundy Island and Clevedon at a time and so we never get to see the whole wave, but we can use the fragments that we do see to piece together the whole wave.

At high water, the crest of the wave fills the channel and at low water we are seeing the wave trough. After high tide and during the ebb, we get to see a part of the slope of the wave running down from the crest to the trough, downwards towards the open sea. During the flood, we see part of the wave slope that runs downwards in the inland direction. The gradient of these sloping parts of the wave is set mostly by the friction on the flow. The surface slope needed to match the friction on a flow peaking at 3 knots (1.5m/s) or so in a water depth of 20 metres (typical values for the Bristol Channel) is about 1cm in 1km. Between the crest of the wave and the trough, the level falls by an amount equal to the tidal range of, say, 6.5 metres, for a mean tide averaged over the channel. Falling 6.5 metres at a gradient of 1cm per km gives a distance between crest and trough of 650km, and therefore the length of the wave – the distance between two crests – is 1,300km. That is a long wave. We can see why only a part of it fits into the Bristol Channel at any time. The profiles on page 157 are snapshots of portions of just one-tenth of the whole wave. The wave speed, equal to the wavelength divided by the period of 12 lunar hours, is about 108km/h. It will take just over an hour for the crest of the wave to travel from Lundy Island to Clevedon, which matches the difference in the time of high tide at these places.

As this wave travels from Lundy to Clevedon, it is squeezed by the converging sides and shoaling seabed of the channel and the resulting concentration of energy increases the tidal range. At the same time, friction removes energy from the wave. Between Lundy and Clevedon, the concentration of energy appears to have the upper hand and the wave grows in size, but only moderately.

In the shallower water upstream of Clevedon, friction becomes more important than the effects of the converging sides and the tidal range reduces.

All this detail of what makes the tide wave work, of course, misses the thrill of what it is like to be out on waters with the power of those in the Bristol Channel. I have worked in a tidal stream of 3 knots (1.5m/s) and that was exhilarating enough; it was like being in a tumbling mountain stream but with the moving water stretching for miles around the ship. The flows in the Bristol Channel can be much faster than that. My friend Alan Elliott kept a boat near Clevedon. He tells me that sailing inland from there up towards the M4 motorway bridge takes you through a region called the Shoots, where the sandbanks converge and the flows are squeezed through a narrow channel no more than 500 metres wide. The Admiralty Chart reports the currents in the Shoots at spring tides as up to 8 knots (4m/s) with 'practically no slack water', which must be close to a record for these islands. It is a white-knuckle ride for sailing boats making this passage, especially as there are rocks on each side of the channel.

The great tides of the Bristol Channel have attracted interest and campaigners of different persuasion because of their potential as a source of renewable energy. Tides have two kinds of mechanical energy: potential energy created as the tide rises and kinetic energy contained in the currents. The potential energy created as the surface between Lundy and Clevedon is raised by 6.5 metres on an ordinary tide is about 10^{15} joules. If that energy could be captured as the tide rises twice each day, the power generated would be about 10^{10} watts. That's a significant proportion of the UK's current energy demand. The most efficient way to capture the potential energy in the tide is to build a dam, or a tidal barrage. Water is allowed to flow freely through the barrage into the upper estuary on the rising tide and then is forced to flow back out through turbines attached to dynamos as the tide falls. Such an arrangement can never capture all the potential energy in the tide: the resistance in the turbines means that the tide falls more slowly inside the dam than outside, and the full range of the tide is not experienced inside the dam. Nevertheless, it has been estimated that a tidal barrage in the Severn could provide something like 5 per cent of the electricity used in the United Kingdom.

Constructing a tidal barrage across the Bristol Channel would be a very large and costly engineering project and there are other problems associated with such a scheme. The electricity generated would cost more than we are currently used to paying (although that could change as fossil fuels run out). There would also be an environmental impact. The tidal range inside the dam would be reduced: it would effectively be close to the current high-water level for most of the time. The sand and mud flats that are currently exposed at low water, and which are important feeding grounds for migrating seabirds, would be lost. There may also be a problem with silting inside the dam and this would affect the power that could be generated.

There are smaller, less ambitious, alternatives to a large tidal barrage, which would generate less electricity but would cost less and have a smaller impact on the environment. Small barrages can be constructed across a bay, for example, to make a tidal lagoon. The flows in and out of the lagoon can be harnessed to generate electricity and local facilities such as a marine walk can be made along the top of the barrage. Alternatively (or in addition), turbines can be placed directly in fast offshore tidal flows and connected to the national grid with submarine cables. These would be the equivalent of windmills on land. Winds are generally faster than tidal flows, but water is much denser than air. A tidal flow of 3 knots has the potential to generate the same electrical power as a wind speed of 60 miles per hour. The big attraction of tidal power, compared to wind power, is that it is regular and predictable. We *know* that the tides are going to continue to flow in and out of the Bristol Channel at speed, twice a day, for a long time into the future.

CHAPTER **7**

INSPECTING THE 'EAGRE'

ATIDAL BORE SWEEPING UP A river is one of the great sights of nature. A bore is the leading edge of the incoming tide; in some rivers it can take the form of a sudden turbulent jump in water level (an attraction for surfers) or it can be a series of gentler waves. Before the bore arrives, the river is flowing sedately to sea; after the bore has passed there is a fast, tumbling flow of water inland and the river level rises rapidly. The world's greatest river bores are destructive affairs. There are accounts of the 'Silver Dragon' on the Qiantang River in China dragging ships off their moorings and leaving their anchors 'shining like they have been newly polished'. The largest bore in the British

Isles, on the Severn, is not so destructive but still demands respect.

There are probably a score or more river bores to be seen on our islands; it is quite probable that there are a few still to be discovered. Bore-watching is an area of marine science in which amateurs equipped with basic equipment and enthusiasm can make an important contribution. Today, this type of contribution would be called 'citizen science', but in the past it was known simply as curiosity.

In September 1928, Mr HH Champion, an inspector of schools, saw the bore on the River Trent in Lincolnshire for the first time. The Trent hosts one of our most accessible tidal bores: there is a minor road running along the river for some distance and the bore can be followed easily by car or – for a fit enthusiast – by bicycle. The bore on the Trent has its own local name – the *Eagre* or *Aeger* – possibly named after a Norse god of the sea, or maybe from the old English word for flood or stream. Whatever the origin, it seems to me that the word 'Eagre' captures the excitement of this phenomenon much better than the word 'bore', which does just the opposite.

After seeing the Eagre, Mr Champion made some enquiries about the formation of river bores in general and this one in particular. When he learned that there had been no systematic study of the phenomenon on the Trent and there was no satisfactory explanation of the cause of river bores in general, he decided to make his own observations. He assembled a team of volunteers, including another schools inspector and his sister, Miss IS Champion, and they set about their work very methodically. The team chose a line of observing sites along the river that they could get to easily by car (a rare resource in the 1920s) and they levelled a temporary benchmark at each site against ordnance datum. A marked pole could then be placed against this benchmark and the river level could be compared from one site to another.

On an observing day, the team would be divided into two or three groups and deployed at observing sites near the mouth of the river (where the Trent joins the Humber), waiting for the Eagre to arrive. They watched for the 'first rise' of the incoming tide and then the group would stay for 10 or 15 minutes, recording the subsequent rise in the river level, before jumping into a car and leapfrogging the other observing groups upriver to the next observing site. In this way, the progress of the Eagre could be followed up the Trent from the mouth to where it ran out of steam at Dunham. Some of the observers would stay at a site after the Eagre had passed and record the water level, at less frequent intervals, for the remainder of the tidal cycle.

The team carried out their observations from 1929 until 1931. Their measurements can be shown as a series of sections of water level during the rising tide, as I have done in the picture on page 167. On a spring tide (there is no bore at neaps), the Eagre first appears near Burton upon Stather and, as it forms, the speed of advance of the 'first rise' of the incoming tide increases from about 3 miles per hour to 10 miles per hour. The formation of the Eagre

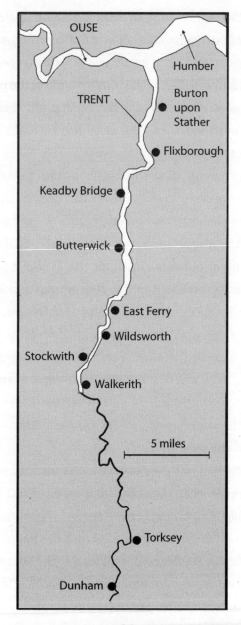

OUSE

Humber

Burton
upon
Stather

TRENT

Flixborough

Keadby Bridge

Butterwick

East Ferry

Wildsworth

Stockwith

Walkerith

5 miles

Torksey

Dunham

accelerates the transmission of the incoming tide upriver. From the riverbank, the Eagre appears as a series of regular waves and it is these that catch the eye, but at the same time as the waves are passing, the river level is also rising quickly. On the scale of the picture I have drawn below, this rise appears as a vertical line, but in actual fact the rise in water level associated with the Eagre takes place over a few minutes, rather than happening instantaneously. As it travels inland, the bore increases in height, reaching a maximum of about 1 metre (3 feet) on an average spring tide at Walkerith and then becoming smaller as it continues to travel upstream.

Mr Champion died in 1931, before he had completed his observations to his own satisfaction. His sister, Miss IS Champion, approached the Liverpool Tidal Institute and

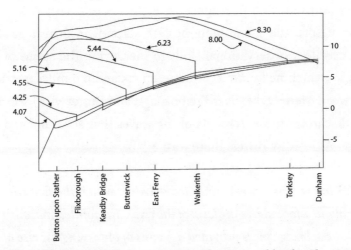

Profiles of the water surface on a spring tide measured by the Champion team. The times by each profile are hours and minutes after low water at Immingham. Figures on the right are heights in feet.

asked if they would be interested in the data and seeing it through to publication. The invitation was accepted and RH Corkan analysed the records and wrote them up, including Mr Champion (but not his sister) as an author. In the published report of the Eagre, the authors draw attention to the steep slope in the river level at low tide between the Trent entrance and Burton upon Stather (which you can see on the extreme left of the figure). This is probably the slope of the riverbed in this section. They wondered if the squashing of the flooding tide at this point might contribute to the formation of the Eagre. We will return to that point later.

The Eagre also drew the attention of an acquaintance of ours from Chapter 4. Vaughan Cornish came to see it in September 1922 and returned in August 1928 (his second visit was just a year before the Champion project started). Cornish applied his excellent observational skills to measuring the speed of the Eagre (which he found to be between 7.5 and 9.9 miles per hour between Mere Dyke, near Flixborough and Butterwick). He also made careful measurements of the waves that accompanied the bore. At Keadby Bridge there was:

'a long train of steep, rounded waves of which the first rose about 1 foot above the level of the river. The jointed masonry of the bridge piers provided a means of observing the rise of the water. One hundred well-marked waves passed the pier in 3 minutes, by which time I judged that the mean level had risen by 1 foot.'

The period of the waves that he saw (the time interval from crest to crest) varied between 1.8 and 2.3 seconds and their height varied along the length of the river. The biggest he estimated as 1.5 metres (5 feet) from crest to trough, at Wildsworth between East Ferry and Walkerith, but the wave height was more usually 0.9–1.2 metres (3 to 4 feet). Cornish also noticed that the largest waves were at the front of the group travelling with the bore. The wave height then decreases moving backwards in the train, rapidly at first and then more slowly so that the 'most spectacular part of the disturbance [was] the first ten or fifteen waves'.

Both Cornish and the Champions reported the presence of a 'second Eagre', which arrives a few minutes after the first. The waves in the second bore are much smaller than the first and it doesn't travel as far up the river as the main bore. I think this second Eagre is a peculiarity of the Trent not seen on other rivers I know that have a tidal bore.

I went to Lincolnshire to watch the Eagre on the big tides in March. The area is reminiscent of Holland: flat and with a sense that the sea is close and being held back by barriers. The River Trent in these parts is enclosed between banks with a good road close by along most of its length and plenty of places to pull over, climb the bank and watch the river. At East Butterwick there is a pub, the Dog & Gun, which seemed a good place to start my investigations. The inside is cosy and the landlord, Joe, was very welcoming. Joe told me that the banks on the river only appeared in the 1950s. Before that the river was allowed to flood over the neighbouring fields and irrigate (and fertilise) them. The old houses all had one foot of clearance beneath them to allow for this: not the new ones, though, said Joe. In the past, the pub had served as a ferry house: the river was crossed at this point. Passengers could have a pint while they waited for the ferry to arrive. The pub on the opposite side of the river (in West Butterwick) is called the Ferry Inn but was closed when I visited and was not regularly open, Joe said.

The following evening, the Eagre was due at both the Butterwicks at 5.20pm and I arrived ridiculously early, fearing I would miss it. I sat on the top of the bank, shivering in the cold, and when the bore did arrive, I almost missed it. It was, at this point in the river and on this tide (not the biggest of the month), very subtle: a rocking of the water surface as much in the cross-river direction as along it. Floating objects had started to move upstream, though, so the tide had turned. It was 5.19pm.

I drove to the Jenny Wren at Susworth, 4km (2½ miles) upriver, expecting to see something similar but I was wrong about that. There wasn't such a panoramic view of the river here: the approach of the bore was obscured by a bend in the river, but I heard the Eagre before I saw it. It was roaring. At first I thought this might be the sound of an approaching lorry but then I saw the bore coming round the river bend. The Eagre now had the clear form of an advancing wave train and the roar was coming from the waves breaking on the far bank. I chased the bore in the car past East Ferry and Wildsworth, where it was getting bigger and the noise was now even greater – sounding more like an express train passing through a station. The height of the waves was biggest at Stockwith. At all the places I saw it, the Eagre marked the transition from calm water flowing seaward to a wave train with a height of 2 or 3 feet followed by turbulent water. At each bend in the river, the waves tried to carry on in a straight line and bounced obliquely off the shore, adding to the general confusion.

The Eagre arrived at Walkerith shortly after 6pm, definitely smaller than it had been at Stockwith, and it was now getting dark. It was time to abandon the chase and retire to the Dog & Gun to write up my notes.

OBSERVATIONS OF THE EAGRE, 21/3/2019

Time	Place	Notes
1719	Butterwick	Almost invisible
1730	Jenny Wren	Heard it coming!
1740	East Ferry	
1748	Wildsworth	
1800	Stockwith	Bore waves biggest here
1805	Walkerith	

In the 1920s, the observations of Cornish and of Champion and his crew showed that the Eagre took around an hour to travel from Butterwick to Walkerith, at a speed of about 10 miles per hour (16km/h). This evening it had travelled faster – making the trip in 45 minutes at a speed of over 12 miles per hour (19km/h). The increased speed of the Eagre when I saw it could have been due to transient conditions (the speed of a bore depends on the river level and so on how much rain there has been in the preceding days). But it is also possible that the bore is faster today because those banks constructed in the 1950s have made the river deeper. The possibility that I might have discovered something about the Eagre and had a plausible explanation cheered me up. I had earned another pint.

The Eagre on the Trent.

The observations of the Champions and their friends and the subsequent analysis by Corkan came to the attention of Arthur Thomas Doodson, the associate director of the Liverpool Tidal Institute. Doodson (1890–1968) is one of the great figures of 20th-century British oceanography. He worked at the Tidal Institute, on Bidston Hill on the Wirral, at a time when Britain had a large merchant navy trading between ports across the globe. Tide tables were needed to allow ships to navigate their way safely in and out of harbour and much of the work of producing these tables was carried out at Bidston. Observations of sea level came into the Institute and were distilled into a few tidal constants – a set of numbers giving the amplitude and timing of the different tidal rhythms – that allowed future tides at the port to be predicted. The working up of the data was carried out by 'computors': people – mostly women – who spent their office hours working up hourly observations of sea level at ports from Aden to Zanzibar (and most others in between) in a way that allowed the tidal constants to be calculated.

Tidal predictions were then made on a tide-predicting machine – a wonderful creation of brass and polished wood – which could be set up to generate tidal curves for a port once the tidal constants were known. Doodson and a small team of his computors famously made the tidal predictions for the D-Day landings on the beaches of Normandy in 1944. He was supplied with the necessary tidal constants by the Admiralty. Their location was, of course, not supplied, although many years later Doodson admitted that he had a shrewd idea about where it might be. It is still not clear how the Admiralty got hold of tidal constants for isolated beaches in occupied France. They probably took the values from nearby ports (known from before the war) and adjusted them using the few observations of sea level available from reconnaissance missions to the beaches themselves. In any event, the predictions were good enough for the job.

Doodson was interested in all things tidal and when he saw the observations of the Eagre on the Trent, he asked two questions that cut right to the core of the problem. The first was: why should there be a bore at all? Why should the water level rise so abruptly when it could achieve the same increase in a gentle way over a longer interval of time? The second question was: why the Trent? What was special about this river that made it have a tidal bore when other rivers, apparently similar, did not? Doodson provided his answers to these questions in his book, *Admiralty Manual of Tides* (written with HD Warburg).

The answer to the first of these questions has to do with the surprising behaviour of flowing water when it is fast and shallow enough. This can happen in the open sea but to keep things relevant to a tidal bore, we will imagine water that is confined between riverbanks. When the flow is steady, not changing with time, the discharge must be constant along the length of the river. The discharge is the volume of water that passes a fixed point in a set time. If the discharge is not constant, water will build up in (or flow out of) some parts of the river and that would not be sustainable. If we simplify matters by thinking of a river that has a fixed width, then the discharge is proportional to the river flow speed v multiplied by the water depth d. If the river gets shallow at a point and so d gets less, then the flow must speed up to maintain a constant discharge and vice versa. I have wondered if this is the physical origin of the proverb *still waters run deep*; it certainly fits the pattern. We can draw a graph of water depth against flow speed and the line of constant discharge will form a curve on the graph as I have shown in the picture on page 176. The flow speed and depth can change along the river but their values must always lie somewhere on this curve as long as the discharge remains constant.

Moving water also has energy. In fact, it has two kinds of energy: kinetic energy, which depends on the square of its speed, and potential energy, which depends on the height of its centre of gravity. When a stream is flowing along a horizontal bed, the height of the centre of gravity just depends on the depth of the water. We can draw another line on the figure that represents the energy of the flow; this is also curved but in the opposite sense to the discharge curve: one is concave, the other is convex. At a point in a river where the flow has a particular discharge and energy, the depth and speed must lie on both these curves; this fixes their values at the points where the curves cross. Remarkably, the curves can cross at *two places*; it is possible for the flow to exist in two states, each of which has the right energy and discharge. One state is fast and shallow (and is called super-critical by river engineers). The other state is slower and deeper and is called, not surprisingly, sub-critical. So long as the flow is over a level bed and the energy and discharge remain constant, these crossing points of the curves are the only possible states for the flow. Other depths and flow speeds do not have the right energy and discharge; they are not allowed.

The two possible flow states are separated by a critical velocity, which is the speed at which waves can travel on the water surface. In shallow water, this critical velocity just depends on the water depth and is equal to $\sqrt{(gd)}$, where g is the acceleration due to gravity and d is the depth of the flow. I've added a line representing the critical velocity to the diagram. Super-critical flow is always moving faster than the critical velocity and that means that waves on the surface cannot make headway against

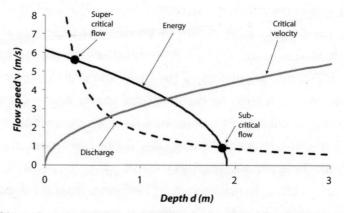

A flow with a given energy and discharge can exist in two different states, a sub-critical one, which is deep and slow, and a super-critical one, which is shallow and fast.

it; they are all swept downstream. Sub-critical flow always moves slower than the critical velocity and waves can travel both up- and downstream.

A pleasant way to watch water making the transition between its two possible flow states is on a walk along the banks of a fast-moving stream. When the water flows over the top of a submerged boulder, it speeds up to maintain the right discharge. If the boulder constricts the flow enough, the flow can reach super-critical speed. Because no waves can travel at the super-critical speed (or even hold their position), the water over the boulder is moving in a rapid flow with a smooth surface.

At some point downstream of the boulder, the flow slows down again (because of friction, if for no other reason). Half a metre or so from the boulder, it will make the transition back to the sub-critical state. If the river floor here is fairly level, this transition will happen as a sudden jump between super-critical and sub-critical flow (intermediate states over a horizontal floor, remember, are not possible). The flow quickly slows down and the surface quickly rises to make the water deeper. The slope of the water surface at this point creates a pressure gradient force, which acts as a brake on the flow, decelerating it from super- to sub-critical speed.

The sudden rise in the water surface is called a *hydraulic jump*. Energy is lost in the jump, some of it to make turbulence and some to make the lovely bubbling sound of the stream. If the loss of energy is small, the water surface after the jump will rise almost as high as it was before the flow arrived at the boulder. If the loss of energy in the jump is great, the level afterwards will be noticeably lower than before. The nature of the jump changes with the flow conditions; sometimes it is a white wall of breaking water; at other times it is a steep but smooth rise in level followed by a series of waves. The water flows through the jump, as you can see by watching the passage of leaves and twigs on the surface. The jump itself is fixed in position relative to the boulder. It can hold this position because it is located right at the point where the flow speed equals the critical velocity. The wavy shape of the jump is travelling upstream at exactly the same rate as the water is moving downstream.

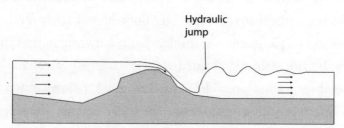

A hydraulic jump behind a boulder in a stream. The flow over the top of the boulder is super-critical and the flow downstream of the jump is sub-critical.

Hydraulic jumps in the majority of streams are small, a few tens of centimetres at most. Occasionally they become much larger, though, and then they really grab the attention. Near our home there is a narrow sea strait through which the tide flows very fast. A line of rocks, called the Bitches, stretches across the strait and, at times of peak current, the squeezing of the flow through the gaps between the boulders accelerates it to super-critical speeds. Some distance downstream the flow jumps back into a sub-critical state in the form of a series of waves that are fixed in position with the water flowing under them. These stationary waves are a great draw to kayakers, who are able to hold their craft in position on the face of the wave. Their weight, pulling them down the sloping water surface, is balanced by the drag of the water on the hull, which is pushing up the slope. The attraction of the hydraulic jump (compared to a surfing wave approaching a beach) is that the waves are fixed in position for an hour or more, giving the kayakers plenty of time to enjoy the ride.

A tidal bore is a *moving* hydraulic jump. The easiest way to see how this can be so and to make the link with the fixed hydraulic jump in a river is to imagine that we are travelling along with the bore, hovering above as it moves at its stately 12mph or so up the River Trent. From this point of view, the river water in front appears to be advancing towards us at a speed equal to the speed of the river heading downstream *plus* the speed of the bore moving upstream. Relative to the bore (and us), the river water is moving fast and at super-critical speed. Behind us, the tide is coming in but the flow speed is not as great as the speed of the bore. From our position travelling inland with the bore, this water seems to be left behind. It is moving away from us at a speed equal to the speed of the bore *minus* the speed of the flooding current. The flow behind is in the sub-critical state. As the water flows beneath us through the bore, it decelerates from super-critical to sub-critical speed and a hydraulic jump is needed to make it change between states.

Here, then, we have the answers to Arthur Doodson's two questions. The water level rises abruptly in a tidal bore, rather than gradually over a longer time, because it is making a jump between the two (and only two) possible flow states that have the right energy and discharge. To create a hydraulic jump, the speed at which the leading edge of the incoming tide meets the river flowing in the opposite direction must exceed the critical velocity – the speed at which waves can travel on the surface of the river. If this condition is not met, no bore will form. By the 'leading edge' of the incoming tide, I mean the point at which the water level is first seen to rise with the tide or, alternatively, the point at which the ebbing current turns to flood (in most tidal rivers, these two definitions will coincide). I think we can agree that the speed of the leading edge of the incoming tide will depend on local conditions – the depth of water, the slope of the riverbed, the tidal range at the mouth of the estuary and so on. Not every river will have a tidal bore. The conditions required to make them are, in fact, rather special and this makes tidal bores rare and something to be treasured.

Doodson thought that the particular shape of the bed in the Trent was important to the formation of the Eagre. The steep rise of the bed near the junction of the Trent with the Humber at first holds back the rising tide and then releases it into the upstream part of the river just as it is reaching mid-tide and is rising at its fastest. The state of the river before the bore arrives is also important in bore formation, although its role is ambiguous. On the one hand, a fast-flowing river will increase the speed at which the river meets the incoming tide and make it more likely that a bore will form. On the other hand, the river may be deep after a lot of rain; that will increase the critical velocity and make the formation of a bore less likely. The depth of the river also affects the speed of the bore and so its time of arrival at a viewing point.

One tool we have today that was not available to Doodson is that computers – electronic ones – can be used to solve the equations governing the motion of the tide filling and emptying a river. This can be done on a regular laptop, which can be used to show an animation of how the water surface changes with time over a tidal cycle. I set up a river with a steep slope to the bed near the mouth and a gentler slope inland, to represent – roughly – the conditions in the Trent. With a tide rising and falling at the mouth of this computerised river, a bore formed on the gentler inland slope in just about the right place, got larger for a while as it travelled inland and then smaller again as it approached the end of its run. The speed of the bore was about right and I thought that I could even see a hint of the 'second bore' that the Champions and Vaughan Cornish reported. I didn't pursue that, but it would be an interesting thing to do some time.

The study of water flowing in streams and channels is called *hydraulics*. The science of hydraulics developed rapidly during the golden age of canal building in the late 18th and early 19th centuries. Canals revolutionised transport at that time. Steam railways had yet to arrive and the roads were full of potholes. A canal barge pulled by horses offered a smooth, safe and comfortable passage. Because the weight of the barge and its cargo was supported by fluid water, much heavier loads could be transported by canal than by road. For passengers travelling between towns, the journey by canal was very nearly as fast, a lot smoother and decidedly cheaper than by stagecoach.

A canal barge, or indeed any boat in shallow water, is influenced by the critical speed – the speed at which waves can move over the surface of the water. A boat pushes the water in its path out of the way, creating a bow wave. The speed with which this wave can get out of the way of the advancing boat is limited to the speed of a wave in shallow water. As we have seen before, the wave speed in shallow water is $\sqrt{(gd)}$, where g is the acceleration due to gravity and d is the water depth. In a canal 2 metres (6½ feet) deep, this speed is 4m/s or about 9mph. As the boat approaches the critical speed, the bow wave cannot get out of the way fast enough (especially in a narrow canal, where it is difficult to escape sideways) and it grows higher in front of the boat, taking its energy from the boat's motion. The boat is trying to climb a steep hill and the horse (or engine) has to work hard to keep it going.

The bargees of the 18th century had discovered how to make their boats travel faster than the critical speed. The trick is to get the horses to put on a short sprint before the bow wave had time to develop properly. This lifted the bows on to the top of the wave and the motion became easy and fast. We would say now that the boat was put *on the plane*. The technique is described in the novel *Hornblower and the Atropos*, in a passage about a journey on the Thames and Severn Canal in 1805:

> *The rhythmic sound of the hoofs of the cantering tow horses accentuated the smoothness of the travel; the boat itself made hardly a sound as it slid along over the surface of the water – Hornblower noticed that the boatmen had the trick of lifting the bows, by a sudden acceleration, on to the*

crest of the bow wave raised by her passage, and retaining them there. This reduced the turbulence in the canal to a minimum; it was only when he looked aft that he could see, far back, the reeds at the banks bowing and straightening again long after they had gone by. It was this trick that made their fantastic speed possible. The cantering horses maintained their nine miles an hour, being changed every half hour.

Putting a boat on to 'the plane' in this way and so allowing it to travel faster than the speed at which waves can travel is the watery equivalent of an aeroplane breaking the sound barrier when it moves faster than sound waves in air, or – in science fiction – of a spacecraft making the jump into hyperspace and travelling faster than light.

Tidal bores, large and small, are always an exciting sight and, although many more rivers don't have them than do, there are plenty of places to see them on these islands. I have listed the bores I know about in the table, although a list like this is out of date almost as soon as it is made. I'm sure there are small bores (forming, perhaps, only on very large tides and rare river flow conditions) to be added. There are certainly some former bores that have disappeared when the shape of a river was altered by flood defences. Some on this list I have seen; in other cases, I am relying on hearsay, or videos on YouTube (some posted by the excellent Rob Bridges).

River	Location
Cree	Dumfries and Galloway
Nith	Dumfries and Galloway
Eden	Cumbria
Kent	Cumbria
Duddon	Cumbria
Lune	Lancashire
Wyre	Lancashire
Ribble	Lancashire
Douglas	Lancashire
Mersey	Near Warrington
Dee	Near Saltney, Flintshire
Taf	Carmarthenshire
Usk	Monmouthshire
Severn	Gloucestershire
Parrett	Somerset
Great Ouse	Norfolk
Trent	Lincolnshire
Shannon	Limerick

The best time to go and see a bore is on a spring tide, a day or so after the new or full moon. Many will only put in an appearance on the largest spring tides of the year at the equinoxes in March and September. Getting the *time* of your visit right is critical and tricky. For the Severn and the Trent, predicted times (and a star system for the likely bore height) are posted on the internet. For other locations, you will have to make your own estimate. The bore is the leading edge of the incoming tide travelling up the river. A starting point in your calculations is the time of low water at the mouth of the river. The advancing tide will then take some time to travel to where the bore first forms. There will likely be trial and error in setting up a timetable and there will be unpredictable variations in time caused by the weather. The golden rule is to arrive early.

The greatest tidal bore in these islands occurs on the River Severn between Sharpness, where it forms, and Maisemore Weir above Gloucester, where it ends. When the family were young, we took a weekend trip to Gloucestershire to see this wonder of nature, staying at a guest house near the river at Minsterworth. At breakfast in most guest houses there are as many conversations as there are tables but this occasion, I remember still, was different. The guests, although strangers to each other, all had something in common. They had come to see the Severn bore, and the conversation was about the best time to set out to the river and the best viewpoint to choose. Advice was passed from table to table. An hour or so later, we found ourselves in a small crowd lining the bank of the river waiting for the bore to arrive. It appeared first at some distance as a silver line, stretched from bank to bank, where the curved face of the leading wave reflected sunlight towards us. As it grew closer, we could hear the noise it made – the rumble – and as it passed the spot where we were standing on a slight curve in the river, it made a great splash, which soaked some unlucky spectators who had got too close. A tall chap with a video camcorder (which, in those days, was the size of a building block) turned to me and said that we had just seen an example of 'concentrated moon power'.

A tidal bore is certainly a demonstration of power squashed into a small space, but the power doesn't come from the moon; it comes instead from the spin of the Earth. The moon creates the tidal forces that pull on the ocean, but it is the Earth spinning within these forces that makes the tide rise and fall. The role of the moon, trying to hold the ocean in place as the Earth spins beneath it, is that of a brake. As a result of this braking action – called *tidal friction* – the spin of the Earth is gradually slowing and the length of the day is slowly increasing. A tidal bore is a small, but very noticeable, example of the sapping of the Earth's spin energy, which is converted in the bore to heat and sound.

We now have several sources of evidence that the Earth is spinning slower than it used to be. The first indications came from observations of solar eclipses, although it was some time before the connection was made between these observations and tidal friction. In 1715, the astronomer Edmond Halley (of comet fame) predicted a solar eclipse that was due to occur in England in that year, using calculations of the motions of the Earth, moon and sun based on the Newtonian theory of gravity. This was the first time that Newton's theory had been used in this way. Halley's predictions were widely advertised in advance and included a map of the path of the eclipse over central and southern England. When the day arrived, the eclipse occurred within four minutes of its expected time. It was a triumph for Halley in particular and humanity in general. Rational thought had finally mastered a phenomenon that had puzzled and frightened people for millennia.

There was a snag, though. After his great success, Halley applied his methods to *historical* eclipses observed by Arabian astronomers in the first millennium BC. This wasn't a straightforward task. There were differences in the calendar to be allowed for and the location of the ancient cities where the eclipses had been observed wasn't always known, exactly. When Halley made the best possible allowances for these difficulties, he found that the eclipse paths he calculated did not pass over the cities where the eclipses had been observed. There was a systematic offset that could only be accounted for if the moon was accelerating in its orbit or the Earth's spin was slowing (or both). It took a further 300 years to solve this particular mystery, with a few red herrings on the way. Tidal friction is slowing the Earth's spin very gradually, and the effect accumulates over time.

As an analogy, if two cars are travelling along side by side and in one the brakes are applied softly, you may not easily spot the difference in speed between the cars but you will notice that one is falling behind the other. The gradual slowing of the Earth's spin causes the position of cities to fall behind the place where they would be if no tidal friction was operating. This happens enough for the hind-casted tracks of eclipses to miss their target.

CHAPTER 8

DOUBLE TIDES
AT PORT ELLEN

ISLAY IS THE SOUTHERNMOST OF THE Hebridean islands and can be reached by the Caledonian MacBrayne ferry service from Kennacraig on the Scottish mainland. For the views alone, the ticket is worth every penny. In each direction there are mountains, islands and headlands with evocative names: Rudha Liath, Lagavulin, Beinn Uraraidh and the Mull of Oa. The place names are a mixture of Gaelic and Norse: a reminder of the time when, in the ninth century, the Vikings colonised this coast and called it the Kingdom of the Isles. By some accounts, this was not a happy time for the resident Celts. They were reduced to second-class citizens and worse until the islands were reclaimed by the Scottish crown in the mid-13th century.

Islay and Port Ellen.

The currents in these waters did their best to harass the invaders. Islay lies close to the northern entrance of the Irish Sea, a body of water with one of the greatest tidal ranges in the world. Each tide, water flows past Islay to fill and empty the Irish Sea. There are two high waters each day and each has a flooding and ebbing current, so that the streams reach their maximum speed four times a day. Peak currents exceed 3 knots (1.5m/s) over a large area extending from Islay to the Kintyre peninsula and the coast of Northern Ireland. The sea around here is fast-flowing and turbulent. Viking longboats powered by sail and oars would have struggled to make progress against the current and they would have been wise to hide in the lee of an island until the flow slackened. Even today, an adverse current slows the ferry to Islay and a favourable one speeds it on its way.

Such tremendous flows over a large area are bound to create interesting effects and one of these is seen in a small area on the south coast of Islay. On most days, Port Ellen (where the ferry from Kennacraig takes you) has an unusual shape to its tidal curve (a plot of water depth against time). I've drawn an example on page 194. Each of the high waters on this day has two peaks, separated by a dip in water level. The *double high water* (as it is called) is particularly prominent on this example in the afternoon, when there is a first high water just after 2pm and a second one at about 6pm.

Double high waters like this are exceptional, although it is likely that (like tidal river bores) more can be found when you start to have a good look for them. The most famous and best studied example is at Southampton on the south coast of England. The extended period of high water at Southampton produced by the two-peaked tide allowed a longer interval for ships to enter and leave the port and gave it a commercial advantage over rival ports trading with Europe. There is also a double high water at Den Helder in the Netherlands and a double *low* water at Portland in Dorset.

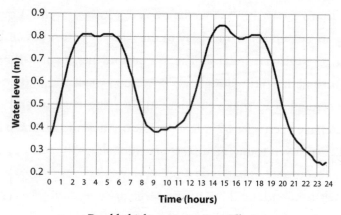

Double high waters at Port Ellen.

My wife, Faith, and I travelled to Port Ellen in a week when the rest of the country was having a heatwave and the Western Isles of Scotland were basked in blessed coolness. We joined the short queue of cars at Kennacraig terminal waiting for the ferry, the *Hebridean Isles*, to arrive. As the boat approached the quay, the bows opened like the jaws of a shark about to engulf its prey. The vehicles that boarded were aligned facing the stern. In the approach to Port Ellen, the ferry turns right round to back in to the harbour stern first so that vehicles can drive straight off. Watching it do this from the shore, in the confined space of the small bay at Port Ellen, you realise that this is a neat bit of seamanship. We were not able to stay long on Islay, but it was long enough to experience an example of how people look out for each other on this island. Driving along a single-track road, we pulled over to allow a farmer to pass. I waved as I would do on the country lanes at home and the farmer – a young woman

Port Ellen lighthouse.

on a quad bike – stopped to check that all was well with us before continuing. You wouldn't get that everywhere.

It is, in fact, difficult to see anything unusual about the tide at Port Ellen with the naked eye. The dip in sea level that happens between the two high waters is small and is easily masked by surface wind waves. It would be possible to stay in Port Ellen for a fortnight or a lifetime and not realise that it possesses special tides. The double high water is a remarkable phenomenon nonetheless. There is a great deal of inertia in the rise and fall of the tide and it takes some considerable effort to stop the level of the sea rising smoothly towards its maximum level and even force it downwards for a while. The impressive thing about the double tide at Port Ellen is not that it grabs the attention – it certainly does not – but that it happens at all.

At Port Ellen (and places like it that have double high or double low waters), circumstances conspire to bring out rhythms in the tide that are hidden elsewhere. Tidal rhythms are the same everywhere on our planet; they are set by the Earth spinning in the gravitational field of the moon and sun. As the Earth turns beneath the moon, two high tides are raised in a period of one lunar day (as we saw in Chapter 6). High tides therefore occur at intervals of half a lunar day, or 12 hours and 25 minutes, on average; we call this the period of the principal lunar *semi-diurnal tide*. A second important rhythm is the *diurnal tide*: a variation in sea level with a period of about one day, produced by the fact that the moon and sun normally lie at an angle (called the declination) to the plane of the Earth's Equator. The diurnal rhythm of the tide tends to raise one of the high (or low) waters in the day relative to the other. In the tidal curve at Port Ellen shown on page 194, the diurnal tide has a high water in the middle of the day. It raises the low water just before noon and lowers the low water around midnight. The diurnal tidal rhythm is a small one for most of Britain and Ireland, but it becomes important at Port Ellen because the semi-diurnal tide is also very small here.

The shapes of the diurnal and semi-diurnal curves on a single day at Port Ellen can be determined by curve fitting. This can be done quite successfully by eye, but I have used a mathematical curve-fitting program to generate the curves in the figure on page 197. The observed tidal curve at Port Ellen is shown in black and the best fitting tidal curves as dashed lines. I have fitted two curves, one with a period of half a lunar day (representing the semi-diurnal tide), the other with a period of a whole lunar day (the diurnal tide). The mathematical name for the shape of these curves is a *sine* wave. When the sum of the best-fitting semi-diurnal and diurnal sine waves is subtracted from the observed tide at Port Ellen, there is a residual oscillation (shown as the grey curve on the figure) that has six high waters during the day; it's called a *sixth-diurnal* tide. It is not very big: the level difference between the highs and lows in this oscillation is less than 20 centimetres, or 8 inches. But then the semi-diurnal and diurnal tides at Port Ellen are not very big either. You will notice that the low waters in the sixth-diurnal tide coincide with the high water in the semi-diurnal tide in both the morning and the afternoon. It is this dip in the sixth-diurnal tide, dragging down the high waters right at their peak, that is producing two humps to the high tide instead of one.

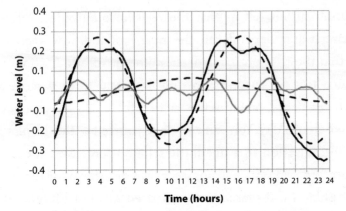

Separating the tide at Port Ellen into its different rhythms.

The sixth-diurnal oscillation in sea level at Port Ellen is an example of something called a *higher harmonic* of the tide. Unlike the semi-diurnal and diurnal rhythms, higher harmonics are not part of the motion of the Earth, moon and sun. They are a feature of tides that is created by the peculiarities of flowing water. The higher tidal harmonics at Islay are probably explained (I think) by the effect of bed friction on the tremendously fast tidal currents in this part of the world. When water flows over the seabed it meets a resistance force, which tends to slow it down. In order to keep it moving, there needs to be a slope on the sea surface providing a pressure head to overcome the effect of friction. As water flows past Islay into the Irish Sea on the flood tide, the water level at Islay sits above that in the Irish Sea in order to provide this pressure head. When the water leaves the Irish Sea on the subsequent ebb, the sea level is lowered at Islay compared to that in the Irish Sea.

Observations tell us that the resistance an object offers to flowing water depends on the *square* of the flow speed (we say that the friction is quadratic). The quadratic friction rule, applied to tidal currents flowing back and forth twice a day, produces a component of the friction force that has a period of one-sixth of a day. You can convince yourself of this by plotting out the variation in a quadratic friction force over a day and separating it into its harmonics, as we did above for the tide at Port Ellen. I did that and it convinced me, but I'll leave it up to you to try if you want. To create the pressure head needed to overcome the effect of quadratic friction on tidal flows, sea level must also have a component that rises and falls six times a day. This is exactly what we are looking for to explain the sixth-diurnal variation of sea level seen at Port Ellen. Most places have a small sixth-diurnal tide, but it is larger than usual at Islay because the currents in the region are so fast and the friction is great.

The effect of tidal friction at Islay (and other places with fast currents extending over a large area) has far-reaching consequences. The rate at which friction removes energy from the tide depends on the product of the water velocity and the friction force. Since the friction force is proportional to the square of the current speed, energy will be lost at a rate that depends on the *cube* of the tidal velocity. A flow of 1m/s will lose energy through friction a million times faster than a flow of 1cm/s. Most energy is lost in places where tidal currents are *fast*: the channel south of Islay is a prime example. It is possible to estimate how quickly the tide is losing energy from observations of currents supplemented by computer models of the tide. Total energy losses through friction in the seas lying on the north-west European shelf are about 10 per cent of the global total, even though these seas have less than 1 per cent of the surface area of the ocean.

The seas of the north-west European shelf are a major sink of tidal energy because they contain large areas of fast currents. To keep the tides going, energy has to be continually fed into them. This energy comes from the Earth spinning within the gravitational field of the moon and sun. The loss of energy through tidal friction is ultimately matched by a loss of energy from the spinning Earth. As we saw in Chapter 7, tides place a brake on the Earth's spin, slowing it down and very gradually increasing the length of the day. The angular momentum of the Earth is passed to the moon, which – every year – moves a little further away. These are slow processes, though. The Earth's spin is a great well of energy from which tidal friction is taking only small sips.

The behaviour of the main, twice-daily tide in the channel between Islay and Northern Ireland (let's call it the Islay Channel) seems quite strange until you realise how it works. The high water sweeps in a circle around the channel in the same way that you can swirl tea around the rim of a cup. The motion can be represented on a map, drawing contour lines through places that have high water at the same time. The lines radiate out from a central point like the spokes of a wheel. At the central point, there is no vertical semi-diurnal tide: it is the point around which the wave swirls. The tidal range – the vertical distance between low and high water – increases outwards from this central point. This pattern is called an amphidromic system (from the Greek *amphi* meaning around and *dromos* meaning running). In the figure below, the times of high water have been labelled in lunar hours after the moon's passage over the Greenwich meridian. It takes 12 lunar hours for the tide to travel around the channel and return to its starting position.

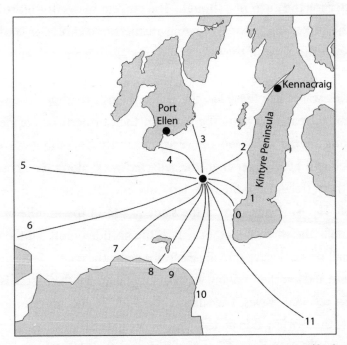

Tidal map of the Islay channel. The numbers refer to the time of high tide, in hours after the moon crosses the Greenwich meridian.

This strange pattern to the tide is created by the interference of two tide waves travelling through the Islay Channel in opposite directions. One wave has travelled from the shelf edge on its way into the Irish Sea and the other has been reflected by the English coast of the Irish Sea and is on its way back out through the Islay Channel. The distance to the reflecting coast and back is such that the two waves interfere *destructively* as they pass through the channel. The crest of one wave arrives at exactly the same time as the trough of the other and vice versa. That is why the rise and fall of the semi-diurnal tide is so small here. It is a different matter for the currents, though. The current beneath the trough of the incoming wave is flowing in the same direction as that beneath the crest of the reflected wave. The currents always *add*: that is why the tidal streams are fast.

Tide waves like these feel the effect of Earth rotation. The wave height increases towards the right (in the northern hemisphere), looking in the direction of wave travel. The incoming wave is therefore a little larger than the reflected wave along the coast of Northern Ireland and the reflected wave is a little larger on the Islay coast. The two waves are of the same height in the middle of the channel. The waves cancel exactly at the no-tide point, but there is a small residual wave moving eastward along the coast of Northern Ireland and another moving westward along the south coast of Islay. These residual waves, travelling either side of the no-tide point, create the swirling tide pattern that we see in the Islay Channel.

Port Ellen is special because it lies sufficiently close to the no-tide point for the higher harmonics to become important *and* the time of high water coincides with the trough, or minimum, in the sixth-diurnal harmonic. It is this coincidence of circumstances that creates the double high water at Port Ellen. Because the sixth-diurnal tide is produced by the effect of friction on the tidal streams (and the streams move back and forth together over the whole region), the phase of the sixth-diurnal harmonic will be much the same throughout the channel. Minima in the sixth-diurnal tide occur every four hours, so somewhere where high tide occurs four hours earlier (or later) than at Port Ellen will also have the right alignment of semi- and sixth-diurnal tides. High tide occurs four hours later than Port Ellen on the coast of Northern Ireland but this is probably too far from the no-tide point for the higher harmonic to have a noticeable effect. On the west coast of the Kintyre peninsula – near the Mull of Kintyre, in fact – the semi-diurnal tides are small and the timing of the sixth-diurnal tide is right to produce a double high tide. As far as I know, no one has found a double high water here, but it would be worth a look.

The double tides on the south coast of England are produced in a similar way to those at Port Ellen – namely by higher harmonics of the tide imposing themselves on a small semi-diurnal tide. There is a wrong-headed view, still sometimes expressed today, that the double high water at Southampton depends on the presence of the Isle of Wight. According to this idea, the tide wave divides as it travels around the island to get to Southampton, one part of the wave travelling westwards and the other eastwards around the island. When the waves meet up at Southampton, one high tide arrives a little later than the other, hence the double high water. The trouble with this explanation is that it can't work. Adding two waves with the same period but with one wave delayed relative to the other just produces a wave with the same period as the originals. There is no double high water.

The first person to show that the double high water at Southampton was the result of higher harmonics was the Astronomer Royal during most of the Victorian era, George Biddell Airy (1801–1892). In February 1842, Airy asked army engineers, or sappers, engaged in the mapping of Britain for the Ordnance Survey to take some time out from their regular work to make measurements of the tide at Southampton. The sappers used a vertical scale attached to the end of a pier and observed sea level at five-minute intervals for five consecutive days (using a lantern to make observations at night). Airy analysed the results in much the same way that we did for Islay – in fact, he was the first person to separate out the tide into its rhythms in this way. Airy found that there were quarter-diurnal and sixth-diurnal harmonics to the tide at Southampton, each about one-tenth the size of the semi-diurnal tide. He didn't know the origin of these higher harmonics but he could see that they were the cause of the double high waters. The central part of the English Channel is an area of relatively small semi-diurnal tides: the water surface rocks about a pivotal line (a nodal line) across the channel so that it is high tide on one side of the line when it is low tide on the other and vice versa. Southampton lies just to the east of this line and high water in the semi-diurnal tide coincides with a minimum in the sum of the higher harmonics. The effect of the harmonics on the semi-diurnal tide is to produce a double peak in the high water. On the other, westward, side of the nodal line, at Portland, the phasing of the tide is such that the low water in the semi-diurnal tide coincides with a peak in the higher harmonics and a double *low* water is formed instead. The timing and size of the higher harmonics relative to the semi-diurnal tide is crucial in deciding whether a double high water, a double low water or, more likely, neither is seen.

Removing the tide from sea level observations to see what is left can lead to interesting results, the equivalent – in a way – of lifting a stone in a rock pool to see what lies beneath. When I was working in North Wales, the central part of the Menai Strait – the seaway that separates Anglesey from the mainland – was a good place to try out new instruments. The water here is sheltered from the worst storms and instruments could be left on moorings where an eye could be kept on them. A favourite spot was near Plas Newydd, the former home of the Marquess of Anglesey (the first marquess led a famous cavalry charge at Waterloo). This lovely building has gardens that lead down to the shore of the strait and the instruments could be left safely in the shallow waters here.

Some of the instruments had pressure sensors and although it was never the primary objective to measure the variation in water level at Plas Newydd, it was immediately obvious that there was something unusual about the tide here. The tidal curve has a sharply pointed peak, like the top of some of the mountains of the Snowdonia range that are visible from this location. The shape is caused partly by a tidal harmonic with a period of one quarter of a day, but there is something else unusual happening. If we fit semi-diurnal and quarter-diurnal curves (the diurnal tide is very small here and there is no need to include it) and then subtract them to see what is left, there is a residual oscillation, shown in the diagram on page 207. This is a different kind of residual to the one at Port Ellen. At Port Ellen, the sixth-diurnal oscillation is there all the time, but the residual oscillation in the Menai Strait appears for a short while and then disappears. It consists of a small increase in water level just before high tide and a small reduction in level just after high tide. That's all there is to it. It is a temporary oscillation, very small, with just a single peak and trough around the time of every high tide. It is what physicists call a *transient*. A good example of a transient – not so commonly seen now as it used to be – is the swinging of a saloon bar door after a cowboy has pushed it open. These doors were hinged so that their weight naturally carried them to the closed position; inertia would carry them past this point and they would oscillate back and forth for a while until the friction in the hinges stopped the motion.

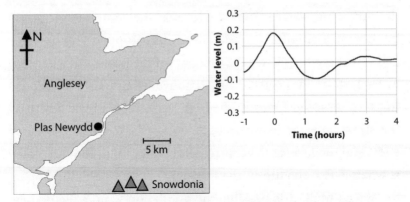

The transient oscillation in the middle of the Menai Strait, visible after the tide is removed.

It is difficult to be sure how the transient in the Menai Strait works exactly because we only have observations in the middle of the strait, but I think that the up-and-down movement of the surface has its greatest amplitude in the centre and is smallest at the ends, like that of a plucked string on a violin. The natural period of such a motion in the strait is two and a half hours, which matches the portion of the transient seen at Plas Newydd.

As the tide rises in the open sea, water flows into the Menai Strait from both ends. The strait is long and narrow and it takes time for the flow to get from the ends to the centre. As a result, the water level in the centre lags behind that at the ends. During the rising tide, there is a slope in the sea surface down towards the centre (the pressure head created by this slope drives the current against bed friction). As high water is approached, the flow slows down and friction is reduced but the inertia in the inflowing water causes the level at the centre of the strait to keep on rising for a while after it has stopped rising at the ends. This is what makes the first part of the transient, when the water level at Plas Newydd rises above what we would expect from a simple tidal curve. The high-water level in the centre of the strait now forces water back outwards, towards the open ends, and the level at the centre falls below that at the ends (this is what causes the dip in the transient). If the tide in the open sea stayed high, it is possible the oscillation would keep going for a while, like the swinging of the saloon bar door, but the currents pick up with the falling tide, friction increases, and the transient is limited to a single up-and-down movement. Each high tide there is a small bobbing movement of the surface in the centre of the strait. The same thing doesn't happen at low tide because friction is greater in the shallower water and quickly damps out any oscillations.

The centre of the Menai Strait is no longer an important seaway and its tide-produced transient is of little practical importance. It is intriguing, nevertheless, to think that it has been happening, quietly every tide, since sea water flooded into the strait after the end of the last ice age. No one noticed it before, because there was no need to measure the tide in this quiet corner unused by large ships. As a result, nature kept this particular secret until it was uncovered by chance measurement.

The tide-made transient oscillation in water level in the centre of the Menai Strait is an example of a wider class of water movement seen in closed, or partly closed, bodies of water such as lakes, bays and harbours. A strong wind blowing over a confined area of sea will pile water up against the weather shore. When the wind stops, the surface tries to level itself, but in a confined space, it will usually overshoot and set up an oscillation in which the water flows back and forth in the space available to it. We have all seen this kind of slopping oscillation in small basins of water. It is very difficult *not* to set one up if you carry a nearly full saucepan across the kitchen. In the simplest form of the oscillation, the surface rocks in a see-saw motion, with the greatest vertical movement at opposite ends of the container and a pivot (a node) in the middle. At the node there is very little vertical movement but it is here that the current flows most rapidly to transfer water from one side of the container to the other. Other types of oscillation with more than one node are possible, but the simplest is the most common.

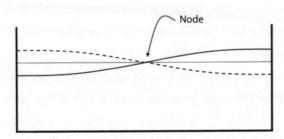

The slopping motion in a basin. The water surface oscillates between the solid and dashed lines.

The period of a single-node oscillation of the water is equal to the time it takes a wave to travel from one end of the basin to the other and back again. This just depends on the dimensions of the basin: its length and depth. The speed of the wave, as we have seen before, is $\sqrt{(gd)}$, where g is the acceleration due to gravity and d the water depth, and the period of a single-node oscillation in a basin of length L is $2L/\sqrt{(gd)}$. For a small saucepan, this period will be a fraction of a second and for a medium-sized harbour or lake it will be several minutes. The period of a two-node oscillation, with nodes one-third and two-thirds the way across the basin, is half the value given by this formula.

The general term for this kind of oscillation is a *seiche*, a local dialect word in Switzerland where they were first investigated on lakes. The pioneer here was François-Alphonse Forel (1841–1912), who established the scientific study of lakes, now called limnology. Forel's home lay on the shores of Lake Geneva and he was familiar with the regular changes in water level that seiches created at the shore. He devised instruments to measure the amplitude and period of the motion and worked out ways to untangle the components of the seiche motion (the lake oscillated both lengthways and widthways and he caught both these signals in his measurements; he also had to separate motions caused by seiches with different numbers of nodes). His work caught the eye of one of the most prominent scientists of the time, George Howard Darwin, son of Charles Darwin. In 1897, Darwin the son delivered a series of public lectures in the United States and his notes were written up in a book *The Tides and Kindred*

Phenomena in the Solar System, suitable for non-specialists. Darwin was particularly impressed with the ingenuity Forel showed in making his own instruments. He wrote:

> *People are nowadays too apt to think that science can only be carried to perfection with elaborate appliances, and yet it is the fact that many of the finest experiments have been made with cardboard, cork and sealing-wax.*

Although I had no experience of making a scientific instrument, I was encouraged by these words to try my hand at making a seiche meter. The most important thing about an instrument to measure seiches is that it must first remove the effect of short-period wind waves, which would be a distraction. This can be done with a stilling well, a vertical tube closed at the bottom but for a small hole to let water in and out. I used a piece of plastic downpipe left over from repairs to a garden shed, stoppered the bottom with a block of wood and drilled a small hole in the side of the pipe. The hole limits the rate at which water can flow from the outside to the inside. The water inside the well cannot follow rapid movements (such as those of waves) on the outside but it can follow the slower rise and fall of sea level associated with seiches and tides. I reckoned that my stilling well would damp out all but the largest waves while leaving any seiche untouched. I finished the job by attaching the bottom of the well to a concrete block and fashioning a float that would find the level of the water inside the well. A rod fixed on top of the float was marked at centimetre intervals. All I had to do was note the level of the rod where it protruded out of the top of the pipe and I had a measure of the water level minus the effect of waves.

Now that I was equipped with a seiche meter, I was keen to try it out. Its first sortie was to a reservoir about 2.5km long in the north–south direction and no more than 500 metres wide at any point. I wasn't sure of the depth of this lake but the period of a seiche is not as sensitive to the depth as it is to the length. If the mean depth is 10 metres, the expected period of a longitudinal, single-node seiche on a lake this size is about four minutes. The wind had been blowing strongly from the south-west for a couple of days and on the day of my trip it had dropped to a gentle breeze, so I hoped these would be ideal conditions for generating a seiche. I carefully placed the meter in the water at the north end of the lake. At first, I took readings of the water level every 30 seconds, then every minute, then two minutes, and then (as it became obvious nothing was going to happen) at five-minute intervals. After an hour, the lake level had stayed stubbornly the same. Although this was disappointing, the time wasn't wasted. It was pleasing to see that the stilling well worked: small waves lapped against the bottom of the well but the float remained steady. An enforced stay by a large stretch of water is nearly always rewarding in some way. At one point a lone bird, large and black, flew in a straight line and at a steady height above the water surface, wings flapping slowly and efficiently. It looked majestic, fearing nothing in its familiar environment and sure of what it was doing. There were also changes to the appearance of the lake to keep me interested. The wind increased noticeably for a few minutes and several windrows formed, about 10 metres apart, running from south to north. They seemed very sensitive to the wind speed, appearing as soon as the wind freshened and disappearing just as quickly when it dropped.

The next outing for the seiche meter was to the sea. Fishguard in north Pembrokeshire is a small town with a first-rate harbour. The approaches to the port are guarded by prominent headlands to the east and west of Fishguard Bay and there is further protection on the northern, seaward, side from a substantial sea wall. The port is now used by the ferry service to Rosslare in Ireland, but it has welcomed grander visitors in its day. For a short while in the early years of the 20th century, Fishguard was an unlikely rival to Southampton and Liverpool in the lucrative transatlantic passenger trade. The Cunard line formed a partnership with Great Western Railways to provide the fastest link between New York and London. Passengers from the United States could disembark at Fishguard and catch the waiting 'Ocean Express' trains to Paddington Station. A number of liners were involved, but the most majestic was the beautiful *Mauretania*, which, for a short while, was the largest liner afloat and, for a longer while, the fastest on the Atlantic crossing. These were heady days for Fishguard. The liners would be met by cheering crowds, some dressed in traditional Welsh costumes. Passengers were ferried ashore on small steamers to two waiting express trains and they could be in London a few hours later. The journey from New York to London could be completed in five days, a record time. When the Great War began in 1914 it became unsafe to cross the Atlantic in a passenger liner and the Fishguard route was not resumed after the war.

There are really two harbours in Fishguard: the deep-water port used by the ferry and the occasional visiting cruise liner and a smaller, picturesque fishing harbour. I arrived at the small harbour one morning an hour before high tide. There was a fresh breeze from the south-east, blowing on my back as I faced the harbour, and the waves were small. The harbourmaster's noticeboard gave the charges for placing things in the water. Since seiche meters weren't explicitly mentioned, I set the instrument up by the side of the boat ramp without paying and began taking readings of the water level at intervals of one minute. It soon became apparent that I might be in luck this time. As the tide rose towards high water, the level in my seiche meter wasn't changing smoothly from one reading to the next but was rising a little and then falling again with a jerky motion.

A pretty harbour on a sunny morning is a popular spot for a walk in the fresh air, and passers-by stopped for a chat and to ask me what I was doing. One couple quickly grasped the idea that the water surface might be rocking within the confines of the harbour and, if so, the period of the rocking motion would likely be just a few minutes. I was impressed by that, but really there wasn't much opportunity to talk because every minute it was time to walk down the ramp to write down another figure.

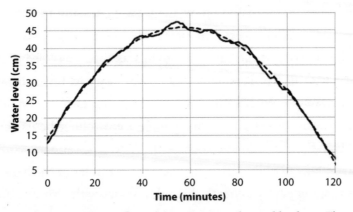

Oscillations on the surface of the water in Fishguard harbour. The continuous line shows the measured water level and the dashed line is a fitted tidal curve.

When I got home, I plotted the measurements out as a graph of water level against time. Over the two hours of the recording, the tide rose towards high water and then fell again but, superimposed on this, there was an oscillation in water level. In fact, there were several small oscillations with different periods; sometimes these added together and sometimes they subtracted so that the amplitude of the seiche changed over the time I was observing it. It looked as though Fishguard harbour was behaving in a similar way to Lake Geneva with seiches sloshing in different directions, or perhaps seiches with different numbers of nodes operating at the same time. The period of the largest oscillation was 16 minutes (which would be a reasonable figure for a single-node seiche operating between the headlands) and the next most important oscillation had a period of eight minutes. The fact that one oscillation had exactly twice the period of the other suggests to me that these sloshing movements were moving in the same direction and that one had a single node and the other two nodes.

The amazing thing really, though, was that I had seen these motions at all. The size of the seiche was tiny – just a centimetre or so – much smaller than the waves in the harbour. While I was making the observations, people were rolling boats down the launching ramp and there was a party of canoeists under instruction, all oblivious to the fact that the water in the harbour was rocking gently beneath them. I, though, with a simple bit of apparatus constructed in a garage, was able to see something that was invisible to the naked eye. It was a bit like building a telescope. That's one of the things I like about scientific discovery: that, for a short while, you can know something (however obscure and unimportant) that no one else on the planet knows. The seiche meter had performed beyond my expectation and it gave me a feeling of quiet satisfaction.

CHAPTER 9

LIGHT AND COLOUR

THE STUDY OF WHAT HAPPENS TO sunlight after it has entered the sea is one of the oldest branches of oceanography, probably because – in the days before sophisticated electronic instruments were available – something could be achieved with just a good pair of eyes. When a boat is in a shallow sea, the clarity or transparency of the water can be determined by looking over the side to see if the bottom is visible. I say 'shallow sea', but there are records of the bottom being visible in surprisingly deep water. In the 17th century, in search of a possible north-east passage (a northern route from Europe to the Pacific Ocean), Captain John Wood in HMS *Speedwell* claimed that he could see shells on the bottom of the sea on the north Russian coast in a depth of 80 fathoms, or 140 metres. For that to happen, light must have travelled down to the bed and back again in straight lines for a distance of nearly 300 metres. That is just possible in the very clearest water.

Usually, though, it is rare that the seabed can be seen from the surface at depths of more than a few metres in the seas on the north-west European shelf. The bottom is obscured by the great number of small particles in suspension in the water. These particles, a mixture of living and dead phytoplankton cells, flakes of clay and other material, scatter the sunlight and prevent it travelling in a straight line for any great distance. The particles are individually tiny – mostly too small to be seen with the naked eye – but together they make the water murky. When the bottom can't be seen, the water clarity can be determined instead by lowering something into the sea from the surface and noting the depth at which the something disappears from view. This experiment was formalised by Pietro Angelo Secchi, an Italian priest and astronomer, who reported (in 1865) his measurements of water transparency made with a white disc. White (or black and white) discs lowered on a marked rope, now called Secchi discs, are still used for making simple and straightforward measurements of the transparency of sea water.

Secchi discs are robust and simple to use. They are great things to get people interested in making their own observations at sea. Sailors on yachts and visitors to a pier can make a unique measurement that can be recorded and added to a database, incrementally increasing our knowledge of how the sea behaves. I am a great fan of the Secchi disc. In my experience, the disappearance depth of the disc (so long as it is not too great) doesn't depend too much on the observer: several people can make their own observations and they will be close. But it is true that a Secchi disc measurement tells us very little about the *way* that sunlight changes with depth in the sea.

Some of the first measurements – anywhere in the world – of the variation of sunlight with depth, and that didn't depend on the human eye, were made in Cawsand Bay near Plymouth in the autumn of 1925. The scientists making these observations, WRG Atkins and HH Poole, used the newly available technology of photoelectric cells, devices that gave an electrical current that depended on the light falling upon them. William Ringrose Gelston Atkins (1884–1959) was head of the department of general physiology at the Plymouth laboratory of the Marine Biological Association. His professional interest was in how the physical and chemical environment of the sea influenced the life in it and he was particularly good at applying emerging technology to measure things in new ways (a skill that is just as valuable today). Atkins knew that the extent to which sunlight was able to penetrate sea water was fundamental to how much photosynthesis could take place in it: in fact, it would place a limit on the biological productivity of the seas and oceans of the entire planet. He approached HH Poole of the Royal Dublin Society, who had been experimenting with photoelectric measurements of underwater light. Theirs was a good match; it produced a partnership that was successful and durable.

Atkins' employer, the Marine Biological Association of the United Kingdom, had been established as a learned body in

1884 against the background of a debate that continues to raise tempers today: fishing quotas. There was a view at the time, held by the eminent biologist Thomas Henry Huxley, that commercial fishing could not destroy the fish stocks of the ocean. We could continue to catch more fish with bigger and better trawlers and the fish would always be there for us. There was no need, according to this view, for quotas. Others were not so sure. One of the loudest voices in support of limiting catches was that of the zoologist Sir Edwin Ray Lankester. Lankester was a younger man than Huxley and was of a new generation of biologists. He had a colourful personality. He was a friend of Karl Marx and it is thought he formed the basis for forthright scientific characters in contemporary novels. Lankester saw the fisheries problem from an ecological viewpoint. The fish that people liked to eat – cod, herring and so on – were not uniformly distributed in the sea but grouped together in shoals; they were vulnerable to targeted fishing and needed protection. It is tempting to think of these two protagonists slugging it out at scientific meetings, dressed in their frock coats like Professors Challenger and Summerlee in Arthur Conan Doyle's novel *The Lost World*, but I'm not sure that ever happened. This was serious business. The British fishing industry was an important component of the largest economy in the world. It provided a livelihood for many and healthy food for many more. It was vital that the industry should be sustainable. Following Lankester's suggestion, the Marine Biological Association was established in the Citadel Hill building next to

Plymouth Hoe (where it still is today) with the aim of promoting and carrying out research into the biology of the sea.

Poole and Atkins faced the problem of making submarine light measurements with a blank instruction book. They were pioneers who had to develop their methods as they went along. It is remarkable, reading their papers today, how modern their work is and how they avoided the potential pitfalls associated with measuring underwater light. Two photoelectric cells were used, one underwater and another on the ship, so that they could allow for variations in surface light as clouds scudded across the sun. The investigators were careful to define what type of light they were measuring. They called it vertical illumination (today we would call it downwelling irradiance). It is the light energy falling on a horizontal surface, regardless of the direction the light falls on that surface. The instrument was fitted with a 'collector' with a diffusing surface (like the ground glass in a bathroom window), which would average the light in this way. Poole and Atkins discussed ways that the illumination on a flat surface was relevant to photosynthesis by a phytoplankton cell, which could collect light equally from all directions. They were also aware that their boat may cast a shadow over the submerged instrument and took precautions against this by lowering from an extended boom over the ship's stern, asking the captain to keep the stern towards the sun if possible.

The measurements were made from the *Salpa*, a converted trawler owned by the Marine Biological Association. The winds were getting up as the autumn set in and the sea in Cawsand

Bay was choppy, with white horses foaming the surface. It was important to let the boat drift so that the drag on the line would be as small as possible and the underwater instrument would hang vertically. In consequence, the boat rolled in the wave troughs and conditions on board would have been uncomfortable. This would not have affected the readings too much, however. The surface photometer was set on gimbals so that it rolled with the ship and the underwater unit was designed to keep the collector horizontal below the surface. There was no pressure sensor on the underwater instrument, so the depth of each measurement was determined from the length of rope paid out with a correction for any deviation of the rope from a vertical line. They figured that the rope would be curved, becoming more vertical towards the bottom, and

The Salpa.

so their correction was quite sophisticated, although small in most cases.

When the ship was on station, the underwater photometer was lowered to the greatest depth at which measurements would be taken and then lifted in increments towards the surface. At each depth, several measurements of underwater and surface illumination were made and an average taken. At the beginning and end of each set of measurements, the two instruments were calibrated against each other by holding the underwater unit above the surface for a while. On the first cruise, five stations were worked over an interval of three days and a sixth station was added a few weeks later. The table below shows the results at their first station in Cawsand Bay: the depth of the observation and the proportion of surface light (as a percentage) that was observed at that depth.

OBSERVATIONS OF UNDERWATER LIGHT IN CAWSAND BAY ON
1 SEPTEMBER 1925.

Depth (metres)	Proportion of surface light (%)
0	100
1.1	63.0
2.1	47.5
4.1	27.5
6.1	15.9
8.1	9.5
10.1	6.4

These figures show the essential features of every profile of underwater light that has been made since. Sunlight in the sea decreases with depth, rapidly at first and then more slowly as the depth increases. This behaviour can be understood by following the path of individual pieces of light energy – photons – as they travel downwards. In water that is uniform in its optical properties, there will be a fixed chance, or probability, of a photon travelling, without being absorbed, through a given thickness of water. For example, if we start with 1,000 photons at the sea surface and 500 are left at a depth z, we know that there is a 50:50 chance of a photon making it through a slab of water of thickness z. We would therefore expect to find just 250 photons left at a depth $2z$, 125 at depth $3z$ and so on. This kind of decrease in the photon count, in which a fixed proportion of the original number of photons is absorbed in travelling a set distance, is called an *exponential* decay. A graph of vertical illumination against depth is a curve with the most rapid decrease in illumination near the surface.

The process by which photons are removed from the downward-travelling sunlight is called absorption. A photon hits a water molecule or a suspended particle and the photon is destroyed: its energy is converted into heat or used in photosynthesis. Photons are also *scattered* by suspended particles. In fact, in the seas around the British Isles, photons are about ten times more likely to be scattered than absorbed; they will bounce from one particle to the next before finally being absorbed. Scattering makes the photons follow a zigzag path instead of a straight line. It increases

224

the distance they travel to reach a given depth and so makes it more likely that they will be absorbed on the way.

One way to express the rate of decay of light in the sea is to observe the depth interval over which half the photons are absorbed: an optical 'half-depth', equivalent to a half-life for radioactive material. For the observations in Cawsand Bay shown in the table, the illumination decreases from 100 per cent at the surface to just under 50 per cent at a depth of 2.1 metres, so the half-depth is about 2 metres. I should explain here that oceanographers are more used to expressing the rate of light decay in the sea using something called an attenuation coefficient. The attenuation coefficient is equal to 0.7 divided by the optical half-depth, but the half-depth is a perfectly fine way of dealing with light transmission through the sea and is certainly more intuitive.

The optical half-depth will vary from place to place and from time to time, depending on the absorption and scattering properties of the water. It is an important quantity in the ecology of the sea. To give an example: as a rough guide, large seaweeds called kelps can survive at a level where the light is about 1 per cent of that experienced at the sea surface. Any deeper than that and it is too dark for them to carry out photosynthesis. Vertical illumination will be reduced to 1 per cent of its surface value at a depth equal to 6.6 times the half-depth. In Cawsand Bay, where the half-depth at the time of Poole and Atkins' measurements was 2 metres, we would not expect to find kelps growing below a depth of about 13 metres (43 feet). (Cawsand Bay is, in fact, known for its eelgrass beds and this rule can be used to plot out the likely distribution of this flowering plant in the bay.)

The tide can be expected to play a role here too. On a day when low water is at noon, there will be more light reaching the seabed (averaged over the day) than on a day when it is high water at noon. As the time of high tide advances by 50 minutes or so each day, this effect will create a fortnightly variation in the daily total sunlight at the seabed. The eelgrass beds in Cawsand Bay (and photosynthetic plants and algae growing in shallow seas with large tides everywhere) should respond to this and show a growth spurt once every two weeks. That would be an interesting idea to test.

While they were making their observations, Atkins and Poole noted the depth at which the light collector on their underwater photometer disappeared from their viewpoint on deck. In their early experiments, they found that the disappearance depth was that at which the illumination had fallen to 30 per cent of its surface value, a result that was, they noted, 'nearly independent of the actual value of the surface illumination over a wide range'. Their collector was smaller and darker than a regular Secchi disc and they later revised their estimate for the disappearance depth of a white object to 25 per cent of surface illumination. The Secchi depth is therefore equivalent to the depth at which illumination falls to a quarter of its surface value, that is two times the half-depth. This provides a handy way of converting easily measured Secchi depths into optical half-depths (or, if your prefer, attenuation coefficients).

The village of Cawsand and its close neighbour Kingsand lie on a small promontory of the south coast of England that is often ignored by visitors on their way to the better-known tourist spots of Cornwall. It is possible to drive there by a roundabout route but an alternative is to take the passenger ferry that runs from the Mayflower steps in Plymouth. On the day that I travelled, a swell coming in from the south-west was gently stirring the water of Plymouth Sound into a brown, cloudy, soup. The deck of the ferry provides good views of the Plymouth waterfront, including the Marine Biological Association building on the hill next to John Smeaton's lighthouse. The ferry runs bow-first onto Cawsand beach and is held there by its engines while the passengers disembark down a sloping gangway onto the sands. The boat's skipper does his best to help but on a day when the sea is choppy it is an interesting challenge to make it to dry land without getting your feet wet.

The ferry makes the return journey to Plymouth after a couple of hours, but I chose to stay the night in a harbourside pub. In the evening it was pleasant to walk along the quiet, narrow village lanes, which looked much the same, I suspect, as they did when Atkins and Poole were lowering their light meter into the sea here. There were a few warships in the harbour that evening, but I was told that these belonged to foreign navies. Things would have been very different in 1925 when Britain had the largest fighting navy on the planet. Warships would have been a common sight to the scientists and crew aboard the *Salpa* as the pioneering light measurements were made in Plymouth Sound.

Natural sunlight comprises light of all the colours of the rainbow, but not all colours are attenuated equally in the sea. Pure water absorbs red light more readily than other colours. It is difficult to see this when the distance involved is small – a tumbler of water, for instance, looks perfectly clear – but the effect becomes noticeable when light has travelled several metres through sea water. The light at depth in a clear sea appears blue as the other colours have been filtered out. Red objects at the bottom of a blue sea look black because there is no red light left to reflect off them. Coastal waters are greener than the ocean because the dissolved and suspended materials that they hold absorb mostly blue light. As sunlight travels down in these waters, the light becomes greener with depth as the blue *and* red colours are removed from the sunlight, leaving just the green.

Unless we dive, we don't normally get to see the colour of the light that is left in the sea at depth. Occasionally, however, circumstances conspire to allow us to experience what it looks like. Once, when installing some scientific equipment in a hut on the end of a pier, I was aware that there was a dim green glow coming up through the floorboards. I turned the hut lights off and closed the door and the inside of the shed was suffused in a vivid emerald green. Sunlight entering the sea at the end of the pier was scattered so that it first travelled horizontally under the pier and then it was scattered again upwards through the floor of the shed. On its journey, the blue and red colours present in the original sunlight had been absorbed, leaving just the green. I was seeing the true colour of this particular bit of sea.

The crew of a ship at sea are interested in the colour of the water because of the information that it can impart. A change from deep blue to greenish-blue, for example, can foretell the approach of land. In high latitudes, a change in sea colour can give warning that ice is ahead. On his voyage to Antarctica at the end of 1914, Sir Ernest Shackleton describes the first encounter with the pack ice:

> *December 7 brought the first check. At six o' clock that morning the sea, which had been green in colour all the previous day suddenly changed to a deep indigo. … Two hours later, fifteen miles north-east of Sanders Island, the* Endurance *was confronted by a belt of heavy pack-ice, half a mile broad and extending north and south.*

I'm not sure why ice surrounds itself with 'deep indigo' waters in this way. My friend David Thomas, who has sailed in Antarctic waters, says he thinks the indigo colour may have come from underneath the ice. Water beneath the ice is devoid of life and so has a very deep blue colour.

A numerical scale for classifying the colour of natural waters was developed in the 19th century by François-Alphonse Forel (whom we met in Chapter 8). Forel used a mixture of chemicals to create a scale of 13 colours, from deep blue to green, which could be compared to the apparent colour of a lake or a sea viewed from the deck of a ship. The scale was later extended by the German limnologist Wilhelm Ule (1861–1940) to include the brown colours of the muddy water encountered near the shore and in an estuary. The colour scale is simple to use and, like the Secchi disc, lends itself to volunteer 'citizen science' programmes. Some care has to be exercised when interpreting sea colour, however. It is easy to be confused by the light that reflects off the water surface (which has the colour of the sky rather than the sea). A good way to view the true colour of the sea is to place one end of a short, straight tube just below the surface and look down the other end, covering it as much as possible with your face. It is also possible to take photographs down the tube and match these to the colour scale.

On an oceanographic research cruise, water colour measurements are made with a radiometer – a light meter that is lowered into the sea looking downwards and which measures the light intensity at a number of wavelengths, or colours. It is possible to reduce these observations to the equivalent colour value on the Forel-Ule colour scale using a mathematical procedure developed by Marcel Wernand of the Royal Netherlands Institute for Sea Research. In some sea areas there are enough observations to draw a contour map of the sea colour, as I have done on page 231 for the Irish Sea. The lower the number on the contour, the bluer the water. Number 4 is a deep blue and number 8 is blue-green. The emerald green I saw filtering up through the floorboards of the pier probably corresponded to number 10 or 11 on the scale. The colour map of the Irish Sea shows features that we would expect in a sea in which physical processes, especially tides, are important. Tidal currents mix mud up from the seabed, making the water opaque and colouring it green. In places where currents are fast,

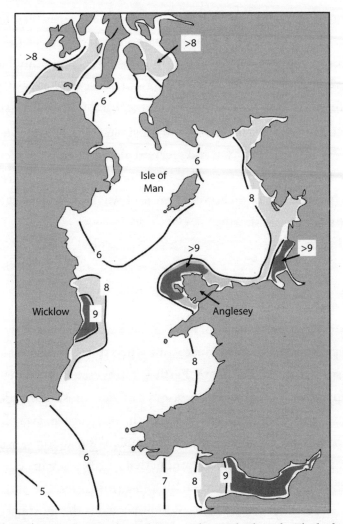

Contours of colour in the Irish Sea on the Forel-Ule scale. The higher the number, the greener the water.

more mud particles can be held in suspension and the water gets greener. Mud, a brown substance, absorbs blue light most, green moderately and red least. In moderate quantities, mud takes the blue colour out of the sea and water removes the red colour, turning the water green. In higher concentrations, such as you might find in an estuary, the mud also absorbs significant amounts of green light and the colour of the water shifts towards khaki. The Irish Sea is clearest and most blue in the deep water to the west of the Isle of Man and greenest along the coast, especially close to the Mersey Estuary and in the Bristol Channel. There are also two patches of green water around Anglesey and Wicklow; their story comes shortly.

There was a great leap forward in knowledge about how the seas and oceans of our planet work when it became possible to measure their colour from Earth-orbiting satellites. Satellites equipped with ocean colour sensors can view most of the globe, cloud cover permitting, every day. In the open ocean, away from the coast, the satellites can see phytoplankton blooms, which tell us about the biological productivity of the ocean. On the continental shelf around the Irish and British Isles, the patterns that the satellites see are instead dominated, for the most part, by the distribution of non-living, or inorganic, particles suspended near the sea surface.

Tiny flakes of mud and clay – inorganic material – are excellent at scattering sunlight. Some of the light is scattered upwards and the sea appears brighter where the water is muddy. Satellite measurements of light scattered from the sea can be calibrated in terms of the concentration of inorganic particles near the surface. Satellite pictures have taught us that the distribution of these particles suspended in the sea around our islands is not at all random; it forms fixed patterns. There are now years of satellite data available and it is possible to draw maps of the long-term average shape of these patterns. CEFAS in Lowestoft have done that (using software developed by Francis Gohin of the French IFREMER marine station in Brittany). I've reproduced an example in the diagram below.

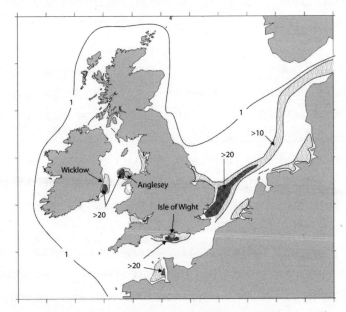

The distribution of inorganic suspended matter at the sea surface in January. Figures are concentration of solids in mg/l.

This map shows the concentration of clay and muddy material, expressed in milligrams of particles per litre of water, in January. The distribution in other months looks much the same, although the concentrations are less in summer than they are in winter. The concentration does not, on the face of it, seem very impressive. A few milligrams of solids in a litre of water adds very little to the weight of the water. But because each particle is tiny, it takes a lot of them to make up the small extra weight. In the areas of highest suspended sediment load, there are close to a million particles suspended in each litre of water. It is not surprising that it is difficult to see things clearly in this kind of water. A photon will travel on average just a few centimetres before hitting a particle and being scattered.

The most prominent feature on the map is the plume of suspended material stretching out from the south-east coast of England. This plume is always present and can be traced across the North Sea as far as Denmark. It is thought to originate in the erosion of the clay cliffs along parts of this coast. The material is then carried, as a suspension of particles, across the North Sea by a current that flows unsteadily but on average towards the north-east, driven by the prevailing south-westerly winds. A plume 20km (12½ miles) wide and 20 metres (66 feet) deep carrying mud at the observed concentration at a speed of a few centimetres per second will transport several million tons of solid material away from the coast each year. The east coast of England is slowly but surely exporting itself into the southern North Sea.

The rapid erosion of the clay cliffs of this part of England (the Holderness coast of East Yorkshire is one of the most badly affected areas) is a real source of concern and worry for many people whose homes and livelihoods are here. It is difficult to know what to do for the best to help them. Putting in shore protection slows down erosion for a while in one area but can speed it up in others. East Yorkshire is not the only area affected and the problem is exacerbated by sea level rise and more frequent bouts of severe weather. It may be that the best way to deal with the incursion of the sea in some areas in the long term is by a managed retreat.

Another notable feature on the map is the patches of muddy water around the Isle of Wight in the English Channel and near Wicklow and Anglesey in the Irish Sea. Each of these *isolated turbidity maxima*, as they have been called, is located in a region of fairly shallow water and very fast tidal currents. The currents are able to stir the mud particles up to the sea surface where they can be seen by the satellite. Unlike the turbid plume stretching from the east coast of England, however, there is no obvious *source* of mud to be suspended at these places. The coast is not noticeably eroding, there are no large rivers and the seabed is composed of pebbles (fine material in the bed has long since been winnowed out by the fast tidal currents). There is a puzzle to be solved here: where is the mud in these isolated patches coming from?

The behaviour of mud in water is a fascinating topic, as every child who has splashed in a puddle knows. One thing a child quickly discovers is that the pool can be made more muddy by jumping into it. Putting energy into the water lifts the solid mud particles from the bottom of the puddle into suspension. Our islands are surrounded by what is, in effect, a large muddy pool and the same principle applies. The energy that holds the mud in suspension is provided, in this case, by tides and winds.

Without an input of energy, mud particles will sink to the seabed under the influence of gravity. The weight of the particle pulls it down and it falls at a steady velocity (called its settling speed) when the weight is balanced by the drag of the water around it. The drag increases with the sinking speed and since the weight depends on the particle size, this balance means that larger particles sink faster. Particles are held in suspension by movements of the water called *turbulence*. Turbulent motion fills the sea and the atmosphere in the form of eddies, or whirlpools, of different sizes and orientation, superimposed on the mean current. In some places the turbulence serves to speed up the mean flow and in others it slows it down. We are familiar with this effect in the atmosphere in the form of gusts in the wind. Aeroplane pilots sometimes warn us that we are about to enter a patch of turbulence in which fluctuations in the speed of the wind will affect the smooth flight of the plane.

Turbulence in the sea transfers parcels of water between adjacent parts of the flow. To see how this can hold particles up in a water column, imagine that the turbulence lifts a parcel of water upwards and, at the same time, another parcel of water is moved downwards; the parcels pass each other as they make their move. Both parcels are then mixed in with their new surroundings. Equal volumes of water move up and down, but since the water moving upwards has come from the deep, it carries more suspended particles than the water moving down. There is a therefore a net movement of particles upwards in the turbulent eddies. In equilibrium, the particle concentration increases downwards at just the right rate for the turbulent transport upwards to exactly balance the sinking. The transport of material, by turbulent eddies, from a place of high concentration to one of low concentration is called *turbulent diffusion*. Turbulent diffusion is a very effective way of mixing material in the sea and in the atmosphere.

The turbulence does something else to the particles it holds in suspension. As they are carried along in turbulent eddies and sink downwards, the particles sometimes collide and stick together to form aggregates or flocs. Aggregates contain many particles of different types held together by organic 'glue'. The average diameter of an aggregate in the sea might be 1/20th of a millimetre but it is composed of much smaller 'primary' particles of diameter typically 1/200th of a millimetre or less. Because they are much larger, flocs sink faster than the individual particles of which they are composed, and this is one of the reasons why the sea is generally clearer in the summer than it is in the winter. Flocculation is particularly prevalent from March to September, when sticky material released by biological activity helps the particles bind together.

There seems to be an upper limit to the size to which flocs can grow and this, too, is controlled by turbulence. The parcels of water carried past each other by turbulence rub together. The friction created by this rubbing becomes particularly important when the eddies are squeezed into a small space and the water moving in different directions comes into close contact. Turbulence is continually losing energy through friction in the very smallest eddies. The energy is taken from the mean flow; it is first put into the largest eddying motions and then passed down through a series of eddies, each one a little smaller than the one before, until it reaches the very smallest eddies where the energy is used to overcome friction. As the mean flow speeds up and more energy becomes available, the turbulence is able to squeeze into smaller spaces. In a tidal flow of 0.1m/s in water depth of 30 metres, the smallest eddies have a size of about 2mm; this is reduced to about 0.3mm in the same depth and with a tidal flow of 1m/s.

If you watch a turbulent flow (a stream carrying small twigs and leaves is a good place to do this), you will see that different parts of the flow, initially close together, move at different speeds and in different directions. If you were to mark the flow with a patch of dye, different parts of the patch would be pulled in different ways. The patch is stretched out and eventually mixes with the surrounding water. If we could zoom in on a part of the flow that was smaller than the smallest eddy, however, and add a tiny drop of dye to that, the turbulence would not be able to get at that to distort it in the same way. The small patch would be carried along by the mean flow and maybe spin around a little in it, but it would keep its shape. Conditions inside this small non-turbulent length scale are very different to those in the main flow outside.

Flocculated particles are fragile entities that cannot withstand the tugging of a turbulent flow. They are able to grow to the size of the smallest turbulent eddies, but no larger. In places where tidal currents are fast, the largest flocs in suspension are small. Smaller particles have slower settling speeds and this helps the turbulence mix the particles up to the sea surface in places of fast tidal currents. It seems that the flocs can adjust quickly to changes in current speed and turbulence. The mean size of the flocculated particles suspended in a tidal current has been observed to change by a factor of four or more between slack water and maximum flow. Turbulence controls the size as well as the concentration of the particles in suspension.

A moment of discovery in science is a rare and wonderful thing. Oceanographers work at piecing together data collected at sea, looking for patterns to help them understand how the ocean works. Sometimes, the pieces fit together and the pattern becomes complete. On the best of these occasions, the new knowledge explains something that has been puzzling for a while. The insight gained may be momentous or it may be small, but either way it creates a feeling of euphoria, which a scientist can express in different ways (I have a friend in the US who dances around the room at these times). At the moment of discovery, nature has drawn back a curtain and revealed a secret that – until you choose to share it – is known to you alone on the planet.

One day at the beginning of 2002, a young researcher at Bangor University called Katherine Ellis was working up some observations from a cruise. She had measured water currents and the size of the suspended particles on the edge of the large patch of turbid water that extends around the north and west sides of the Anglesey coast. The measurements showed that there was a horizontal gradient in the size of the particles: they became smaller, on average, with distance into the region of high turbidity. This made sense: the particles in the turbidity maximum were being held in suspension by the fast tidal currents and these could be expected to tear up the flocs, making the mean particle size small here. When she combined the particle size measurements with the current observations, however, she found something that she hadn't expected. Averaged over a tide, there was a transport of large particles *into* the turbid zone and a transport of small particles back out again. In the same piece of water, the currents were carrying different sized particles in opposite directions.

Here, at last, was the secret to the source of the material in isolated turbidity maxima. The strong turbulence inside the turbidity maximum tears up flocs, increasing the concentration of small particles and reducing the concentration of large ones. This makes a gradient in the concentration of small particles, decreasing out of the maximum. As we have seen, turbulent diffusion always works to move material down a concentration gradient. Small particles diffuse down their concentration gradient into the water surrounding the maximum. As they do so, they move into an area of weaker tidal streams and are able to flocculate into large particles. This creates a gradient in the concentration of large particles downwards *into* the maximum, in the opposite direction to the concentration gradient of small particles. Turbulent diffusion carries the large flocs down their concentration gradient back into the maximum, where they are torn up into small particles again and the cycle continues.

In this way, isolated turbidity maxima are able to exist without a local source of suspendable matter; instead, they feed on the large flocculated particles in the surrounding water. They suck in flocs and spit them out again as their smaller component pieces. Inside the maximum, the turbulence is great and the particles are small and have slow settling speeds: they are easily mixed up to the surface, where they can be seen from space. If this interpretation is right, isolated turbidity maxima are self-sustaining entities; they will spring up wherever there is a region of fast tidal currents flowing through muddy sea water.

Sea-colour measurements from satellites have taught us other things about our waters (and no doubt will continue to do so as technology develops), but there is room to mention just one more application here. Fresh water flowing from rivers into the sea carries with it a chemical reminder of the land it is leaving behind. Streams and rivers pick up dissolved material, especially when they flow over peaty land, which stains the water brown. You have probably seen this in a mountain stream: a clear amber colouring, rather like tea. The colour is produced by decaying plant material. It is called 'yellow substance' in English and the German word 'gelbstoff' is also commonly used. Its scientific name is coloured dissolved organic matter (or CDOM).

When the fresh water enters the sea, the yellow colour is diluted through mixing. It is no longer visible to the eye but can still be detected with sensitive instruments. If the water is clear of particles (which would otherwise dominate the colour), it can also be picked up by radiometers on satellites. The Clyde Sea, in the south-west corner of Scotland, has weak tidal streams and it is relatively deep. Fresh water flowing into this sea from the River Clyde spreads out on the surface, creating stratification that is present for most of the year. Particles sink to the bottom, leaving clear water at the surface. The Clyde Sea has some of the most transparent waters to be found anywhere around our islands.

As fresh water (mostly from the River Clyde, but also from several smaller estuaries) mixes with salt water in the sea, the intensity of the yellow colour tells us the proportion of fresh water in the mixture. In other words, we can use observations of sea colour to estimate the salinity of the water. To convert the colour into salinity, it is necessary to make measurements at sea of the relationship between gelbstoff concentration and salinity, but that's not a difficult task. There is always an inverse relationship (the gelbstoff decreases as salinity increases) and it is usually a strong one. Once that is established, the colour measurements from space can be converted to maps of surface salinity.

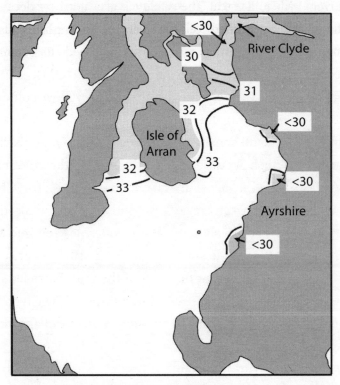

Surface salinity of the Clyde Sea determined from water colour. Salinity is expressed as grams of salt per kilogram of sea water.

A tentative step in this direction was first taken by Caren Binding, a scientist now working for the Canada Centre for Inland Waters. The result is shown in the map on page 243. The changes in sea colour produced by yellow substance flowing in off the land are subtle and are only just detectable, but the salinity patterns that appear agree with what we know about the surface circulation of the Clyde Sea. There is an anticlockwise current around the Isle of Arran, as the Coriolis effect deflects the water emerging from the River Clyde to its right. The satellite is also able to pick out the small low-salinity plumes from the rivers on the Ayrshire coast.

CHAPTER 10

LAYERS IN A LOCH

W E LIVE IN A WORLD OF stratified fluids. The density of the atmosphere, the ocean, and a pint of Guinness all increase downwards, towards the centre of the Earth. This order of things is arranged by the Earth's gravity, which will always pull a dense blob of water downwards through less dense water and push a bubble upwards. In some places – the interior of the deep ocean is one – the gradient of density with depth is gradual. In other cases, especially near a boundary such as the sea surface or the seabed, *mixed layers* of uniform density are formed. One layer lies on top of another, denser, layer separated by an interface across which the density (and other properties of the water) changes rapidly.

Nature occasionally puts on a show to remind us that the atmosphere and ocean are layered. The stripes of parallel clouds seen in the atmosphere are produced by waves on the interface between strata of air of different density. As the air rises to the top of the wave, it is cooled and water vapour condenses into clouds along the parallel crests. Thankfully a thing of the past in this country, the 'pea-souper' fogs that used to occur in industrial towns and cities were caused by smoke from domestic and industrial chimneys being trapped in a layer of dense air close to the ground. The Clean Air Act of 1956 was designed to solve this problem (and eventually it did), but I can still vividly recall being caught in a dense fog in Salford in the late 1960s. It was an eerie experience waiting at the bus stop, watching vehicles loom up out of the greyness. The brightly illuminated numbers on the destination board of the bus only became visible when they were just about in touching distance.

We are less familiar with layering in the sea, although that is just as important – or more so – to marine life than atmospheric stratification is to us. The heating of the sea at the surface creates a warm upper layer separated from deeper water by a rapid vertical gradient of temperature called the thermocline. Near the coast, fresh water running into the sea from the land makes a surface layer of low-salinity water. In each case, light water sits on top of heavy; the situation is gravitationally stable in the same way that a pendulum bob hanging from a string is stable. Like the pendulum bob, stratified water can be set in motion with a push. It is fascinating to watch the movement of layers of water with different density; the motion is slower and more elegant than you might expect. The performance can be created in a glass using fresh and salt water, with a little food dye or milk to colour one of the layers so that it can be seen. A good way to fill the glass without mixing the layers is first to put some fresh water into the glass, up to a little below the halfway mark. The salt water (made with a spoonful or two of table salt mixed into tap water) is then introduced slowly to the bottom of the glass through a funnel with a small piece of cotton wool at the outlet to stop the flow being too fast. The salt water fills the bottom of the glass and pushes the fresh water up.

To create a wave on the interface, tilt the glass over for a moment and then put it straight. For an instant, the interface has a slope as in the picture below, but the sloping interface creates a pressure gradient in the bottom layer that sets the water moving one way in the bottom layer and in the opposite direction in the surface layer. The interface becomes level but then overshoots and oscillates, tilting first one way and then another about a central axis, like a see-saw. The motion is that of a standing wave: a point on the interface moves up and down and the currents oscillate back and forwards, but the wave doesn't travel horizontally. The standing wave trapped in the glass is very similar to the seiche we saw in Chapter 8; the difference is that in this case the wave is not on the water surface but on the density interface. It is called an internal seiche.

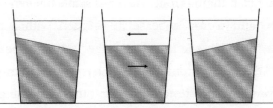

An internal seiche in a glass of fresh and salt water.

When I did this experiment in a half-pint glass, the oscillation of the interface had a period of about one second and kept going for almost a minute. In contrast, the motion of waves on the surface of the water died away in a fraction of a second. The movement of the internal seiche is slower than a wave on the surface because the restoring force depends on the difference in density across the interface. At the water surface the difference in density is that between air and water – about 1,000kg/m³. Across an interface between fresh and salt water, the density difference will be much less (maybe just 1 or 2 kg/m³) and the period of the motion is consequently longer. I calculated that the period of a surface seiche in my glass would be about 0.1 seconds, one-tenth that of the internal seiche that I saw. In a large body of stratified water, such as a lake or lagoon, the period of a surface seiche would be counted in minutes and that of an internal seiche in hours.

The energy in the internal seiche changes from potential to kinetic and back again during the oscillation. At the start of the motion illustrated in the figure on page 247 (left picture), salt water has been lifted on one side of the glass and fresh water lowered by the same distance on the other side. Because the salt water is heavier, there is a net increase in the height of the centre of gravity, and so an increase in potential energy compared to a glass in which the interface is horizontal. One-quarter of a period later, this potential energy has become kinetic energy as the layers of water flow over each other (middle picture). A further quarter period later, the interface is tilted upwards the other way (right picture) and the energy is once again potential, ready to start the flows going back again.

A sharp density interface, like the one in the glass, is a great inhibitor of vertical mixing. This is because energy is needed to overcome the density difference across the interface. If, for example, a portion of the lower layer is pushed up through the interface, it will be heavier than its surroundings and will tend to sink back again, unless enough work is provided to project it irreversibly into the upper layer. Similarly, a piece of buoyant water from the surface layer will tend to rise back upwards if it is pushed down. The sharper the interface, the more of a barrier it becomes. This is just as true in the sea as it is in the glass.

Internal seiches can form wherever water is stratified and contained between at least two coasts so that a standing wave can form, trapped between opposite shores. A good place to study them is in a lake where there is no tide and in which the internal seiche can be the dominant mover of the water. The study of lakes is strictly a different discipline to oceanography, but many of the principles are the same. Lakes become thermally stratified in summer and autumn, with a layer of surface water warmed by the sun lying on top of cool water, the layers separated by a thermocline. Seiches on the surface of lakes were first noticed on Lake Geneva, as we learned in Chapter 8, but internal seiches on the thermocline in a lake were discovered – in the sense that they were first correctly identified – in a classic study in Loch Ness in Scotland.

Loch Ness is a fresh-water lake in the Great Glen; it is long, narrow and straight. The loch is also deep – more than 200 metres (650 feet) for much of its length; that's deeper than most of the seas on the continental shelf around our islands. There's a lot of fresh water in Loch Ness: more than that in all the lakes in England and Wales combined. The seiches that form in the interior of the loch through the summer were first noticed in 1903 by a team making observations of the subsurface temperature of the lake. The team included Sir John Murray, an eminent oceanographer who was already famous for his involvement in publishing the 50-volume scientific reports of the Challenger Expedition – the first oceanographic research cruise. Murray was in his sixties by the time of the Loch Ness survey but he wasn't treating this as a retirement project. He would later embark on a survey of the North Atlantic. Murray established a marine laboratory in Edinburgh in 1884; this later moved to Millport on the Isle of Cumbrae and evolved into what is now the Scottish Association for Marine Science, based at Dunstaffnage, near Oban.

The results of the temperature survey of Loch Ness were written up by ER Watson in 1904. Watson's prose is modern, lucid and engaging. His paper draws the reader into the investigation, pointing out the puzzling nature of the observations and how the problem might be resolved. I was left with the impression that it would have been a privilege to sit with these investigators as they worked up and down the loch in their steam launch *Sunbeam*. As part of their investigation, the team measured a series of vertical profiles of water temperature, once or twice on most days, for several months at a station near Fort Augustus at the south-west end of the loch. They were helped in their work by the Benedictine monks of the abbey at Fort Augustus. The abbey boathouse was used to house electrical recording equipment: the temperature measurements at Fort Augustus were made by a string of platinum resistance thermometers connected by cable to the recorder. This was a very advanced procedure for the time. The equipment proved troublesome but, when it was working, it allowed the investigators to make observations in all weathers and at night.

The observations at Fort Augustus showed that the temperature down to any depth above 100 metres (328 feet) was prone to large and sudden fluctuations that could not be explained by heating and cooling at the surface. In Watson's words:

> *The only conclusion which could be drawn from such observations was that the waters of the lake were continually moving – that there were considerable currents throughout the whole volume of water from the surface right down to 300 feet at least, and that the temperature changes were due to these currents which were continually replacing the water at any one point by water with some other temperature.*

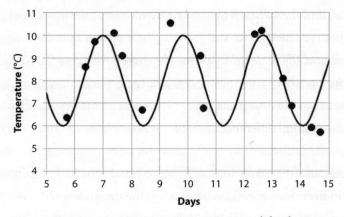

*Temperature measurements at 61 metres' depth
in Loch Ness in October 1903.*

I've plotted out some of the results from the Fort Augustus station above. The graph shows the temperature measured at a fixed depth of 61 metres (200 feet) below the surface over a ten-day period in October 1903. The temperature rises and falls by about 4 degrees Centigrade during this time. The curved line shows a repeating pattern of temperature with a period of 68 hours.

When it comes to interpreting what might be causing these fluctuations, Watson describes a laboratory experiment in which a 'swinging of the interface' of a stratified fluid is produced in a tank. It is essentially the same as the experiment I described in the half-pint glass, but using oil and water to create the layering. Here is Watson's description of the experiment:

> *If we take a long rectangular trough with glass sides, and put
> into it a layer of water, and above the water a layer of lighter*

oil, and then disturb the arrangement, one of the movements observed will be a swinging of the interface between the oil and the water. The longer the trough, the slower will be the period of this movement; a large difference in density between the upper and lower layers will give a quick period, and a small difference of density a slow period.

Connecting this experiment with the observations in Loch Ness was, in my opinion, a wonderful leap of the imagination – a moment of inspiration that at once provided a simple physical explanation for a puzzling set of observations. Watson and his colleagues had seized the opportunity that their observations offered them. They had correctly identified the bits of their data that could not be explained by contemporary thinking. They had appealed to laboratory experiments and theory until they came up with a plausible explanation. The survey had not set out to investigate internal seiches – at the time, no one knew such things existed. If the team had stuck to their original goal of measuring temperature changes, their results would have made a small footnote in history. But because Watson and his colleagues were open-minded and saw something strange in what they measured, they discovered something entirely new. This is a great piece of science.

The internal seiche in a lake is set in motion in the same way as a seiche on the surface of the water. A wind blowing along the lake piles water up at one end and the weight of surface water pushes down the thermocline. When the wind stops, or weakens, the thermocline tries to adjust back to the horizontal, overshoots and an oscillation is set up. In the simplest form of this motion, the thermocline rocks in see-saw fashion about a pivot in the centre of the lake, as in the figure on page 247. As the thermocline rises and falls, the temperature at a fixed depth goes down and up. The period of the temperature fluctuations will be the same as the period of the motion of the thermocline and equal to the period of the seiche. Watson was able to estimate the period expected for the temperature fluctuations from theory, using observations of the depth of the thermocline and the change in temperature across it. This estimate was 68 hours, which is in good agreement with the observations.

The vertical gradient of temperature measured by Watson at a depth of 61 metres in Loch Ness was about 0.1 degrees per metre, so if the fluctuations in temperature of about 4 degrees are to be accounted for by the thermocline moving up and down, the size of this up-and-down movement must have been about 40 metres. Internal seiches in lakes create waves fit for giants. If I had been able to fashion the float of my seiche meter (Chapter 8) so that it sat on the interface of the thermocline in Loch Ness, I would have had no problem in seeing the vertical movements. A seiche of that amplitude will move a volume of nearly half a billion cubic metres of water from one end of the lake to the other every few days. Although we cannot see them from above the surface, internal seiches create large movements within a water body.

Faith and I travelled to Loch Ness on the overnight sleeper train from London. Can there be a better way to travel overland? Well, perhaps there can. We both found it difficult to sleep, but that is perhaps because our bodies are no longer suited to bunk beds. The train arrives at Inverness station in time for breakfast and the former abbey at Fort Augustus is then a short drive to the south-west, past Urquhart Castle along the A82. On the day of our trip, the sun was shining and the reds and golds of the trees around the loch were at their brilliant autumn best; it was a magnificent sight. The abbey at Fort Augustus is now a private holiday park but the boathouse is open to the public as a restaurant. The setting of the boathouse, with its views down the length of the great loch, is spectacular. What a wonderful place this must have been for Watson and his colleagues to visit and work. I wandered around the site absorbing the atmosphere, not expecting to find any connection with the scientific breakthrough that had happened here, but I was about to be surprised and delighted. There is a small cemetery not far from the boathouse, set aside for the monks of the abbey. Each grave is marked by a white metal Celtic cross; there is no regard to status, each cross is identical apart from the inscription. Among the names was one I recognised. Odo Blundell (1860–1943) is one of the monks thanked in the scientific reports for his enthusiastic help with the work. Between September 1904 and April 1905, the scientists had departed to work on other lakes and the temperature measurements in Loch Ness were left in the hands of the monks of the order of St Benedict. Here was a tangible link to the scientific history of our home waters and it lifted my spirits to see it. Blundell's cross now overlooks the loch whose secret he helped to uncover.

At first, there was some scepticism about Watson's explanation of the temperature fluctuations in the interior of Loch Ness. Soon, though, internal seiches were discovered on other lakes worldwide and the doubters faded away. Loch Ness remained a convenient – and iconic – place to study the phenomenon, however. As the years progressed, more advanced (and reliable) instruments revealed features of the internal seiche in the loch that were unknown to the earlier investigators. In 1970, a young researcher from the National Institute of Oceanography, Stephen Thorpe, saw that the motion of the thermocline was not really the 'swinging interface' that Watson had described, but that the temperature at a point first rose rapidly and then fell slowly. Instead of a standing wave, the motion of the thermocline is, on occasion, a surge: a depression travelling along the interface in a similar fashion to a bore travelling on the surface of a tidal river.

The formation of the surge is an example of what is called a non-linear effect, something that happens when the variation in the thickness of the layers along the length of the wave become important. In the picture of an internal seiche on page 247, the thickness of the layers below (or above) the sloping interface changes from one side of the glass to the other. When the seiche starts up, the currents will transport a little more water where the layer in which they are flowing is thicker. This causes the interface to steepen at one point and form into a surge that travels along the interface. The transition is a gradual one: the motion starts as the rocking motion that Watson described and evolves over time into a travelling surge. The internal seiche in Loch Ness lasts for long enough to allow the surge to form. You can make a surge on the surface of water in a shallow trough – a piece of guttering with caps fitted on the ends is ideal. Half fill the guttering with water and then lift one end slightly. When you put the end down again, you will see a train of short-crested waves that follow a sudden rise in water level and travel along the surface, bouncing from one end of the gutter to the other. In Loch Ness, the distance between crests of the train of internal waves is about 1km (⅔ mile). Just like in the gutter, the internal bore and its train of waves travel up and down the loch, bouncing several times off the ends of the loch before running out of energy.

The accounts of temperature measurements in Loch Ness make no reference to the M-word. Indeed, it is quite possible that Watson and his colleague had never heard of the Loch Ness monster; the legend became popular only in the 1930s. Ironically, though, the patterns created by the reflection of sound by the internal motion of the loch did become part of the debate. When echo sounders became available, they were used in the search for large animals in the deep waters of

Loch Ness. Sound bounces off density interfaces in water and the sinuous shape of the surge and its train of waves travelling along the thermocline did sometimes take on the shape of a lake serpent.

The temperature layering of a fresh-water lake is created as the surface waters are warmed by the sun in the spring and summer. In the late autumn, the lake surface starts to give its heat back to the atmosphere and begins to cool. Convection currents form, with cool water sinking from the surface and plunging down into the thermocline. This, together with wind mixing by autumn gales, destroys the temperature layering in the lake. The overturning, or mixing, of thermal stratification at the end of the year is an important part of the ecology of lakes and seas. It reunites bottom and surface waters. Nutrients in the bottom layer are mixed to the surface and oxygen from the surface is mixed down to the bottom.

Oxygen is essential for nearly all life in water – fresh and marine. Oxygen molecules sit in between water molecules, in a ratio of about one or two in a hundred thousand. The oxygen is said to be dissolved. There is usually plenty (in the sense that there is enough for respiration) of dissolved oxygen near the water surface, where it is mixed in from the atmosphere and produced in photosynthesis by marine algae. Near the bottom of a deep stratified lake or sea, however, the oxygen is used up by respiration more quickly than it is mixed down from the surface. Animals will struggle to breathe here in the late summer and autumn. The problem is exacerbated by bacteria decomposing dead material that sinks to the bottom from the surface – a process that also uses up oxygen.

The overturning of the water column in the autumn therefore brings a welcome breath of fresh air to bottom waters, charging up their oxygen levels so that they will be habitable the following summer when the stratification forms again. There are some places, however, where there is no annual overturning; instead, stratification lasts all year, for several years in succession. When this happens, the continued consumption of oxygen in the bottom layer by respiration and bacterial processes can strip it out to such an extent that animals suffocate. The water becomes stagnant and fish kills result.

Stratification that continues for several years can occur when salt water becomes trapped in a deep basin, separated from the sea by a sill, as in the sketch on page 260. This picture could apply to fjords and sea lochs, for example, places that were gouged out of solid rock by ice and which have a shallow sill where they connect to the sea. One of the most extreme examples that I know of is to be found in Lough Furnace in County Mayo on the west coast of Ireland. Lough Furnace is a small lake, about 1km (⅔ mile) wide and 2km (1¼ miles) long, connected to the open sea at Clew Bay by a narrow and shallow channel. Fresh water comes into the lake from the larger Lough Feeagh (which lies further inland and to the north); the fresh water flows over the top of the hollowed-out basin in Lough Furnace and out to sea through the channel. The tides in Lough Furnace are small and mixing is weak; a strong interface is set up between the fresh surface and salty bottom layers in the main northern basin of the lough. The water is not very deep – no more than 20 metres (66 feet) – but the stratification is strong enough to resist the cooling of the surface waters in winter. Bacterial activity in the bottom layer consumes oxygen and a deep pool of deoxygenated water, occupying about a third of the volume of the lough, is formed.

Animals that can swim avoid the deoxygenated waters of the deep layer by inhabiting the shallow margins of the lough, but there is still a problem. The wind creates an internal seiche, which sets the deep, deoxygenated layer in motion. The period of the internal seiche is about two hours and the motion is that of a standing wave: when the interface is rising on one side of the lake, it is falling on the other and vice versa. The motion of the internal seiche appears to be linked to the deaths of European eel living at the edge of the lake. As the interface rises, the deep layer of low-oxygen water creeps towards the shore. During this upwelling of the deep layer, the dissolved oxygen concentration at the edge of the lough falls from safe to dangerous levels. European eels are pretty good at coping with low oxygen levels for a short while, but the regular dose of

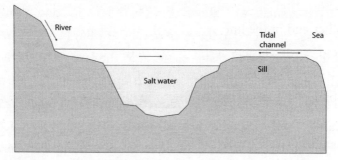

Salt water trapped in a deep basin separated from the sea by a sill.

suffocating waters, repeating every two hours, is too much for some of them. The deep deoxygenated layer is a menace to all the animals living in Lough Furnace, especially the ones that are not nimble enough to avoid the movements of the internal seiche.

⁓

The west coast of Ireland can be reached by train from Dublin's busy Heuston station. While waiting for the train, you can try a pint or two of Guinness in the excellent station buffet. Guinness in Ireland is served differently to beer in a British pub. The bartender will partly fill your glass and then leave it to stand while they go to serve another customer. At this point, overseas visitors should avoid the temptation to shout 'hold on, you haven't finished my drink yet'. That would go down badly and possibly start a fight. Be patient and after a minute the bartender will return and finish pouring your drink. If you have behaved yourself, they may even finish it off with a clover-shaped flourish on top of the white foam.

The train journey passes through the heart of Ireland. Galway, at the other end of the line, is a prosperous and popular destination. The old town clusters around the harbour and estuary of the River Corrib, but it is surrounded by a much larger hinterland of new-builds. The world and his aunt want to live in Galway and local estate agents set their prices accordingly. Continuing the journey to Lough Furnace involves hiring a car and driving north, along the coast road known as the Wild Atlantic Way. You pass through

Clifden, near where John Alcock and Arthur Brown landed after making the first non-stop flight of the Atlantic on 14 and 15 June 1919. They took off from St. John's in Newfoundland in a modified Vickers Vimy and landed for what they had planned as a brief stop in Ireland before continuing to England. The wheels of the plane got caught in the soft ground when they touched down; the still-turning propeller plunged into the bog and that was the end of their journey. Nevertheless, they had crossed the ocean in a time of 16 hours, in a cockpit open to the elements. It was a great achievement that cheered the country at a time when it was coming to terms with the losses of the Great War and suffering from a serious flu epidemic.

County Mayo is well served by walking trails, one of which makes a circuit of Lough Furnace. Derradda Community Centre, where there is a small car park, makes a convenient starting point for the walk. It was here that I learned that the lough gets its name from the iron works that used to be nearby. As I set out, the sun was shining and clear amber-coloured streams charged with overnight rain were tumbling their way into the lough. A couple of workers at the Marine Institute's fisheries laboratory on the northern side of the lough chatted in Gaelic. It was about here that it started raining hard and I decided to sit down in some shelter and put on a pair of waterproof trousers. It was pleasant sitting under the trees, watching the rain slanting down. I was trying to remember the trick of getting overtrousers over walking boots using only a body that falls over if it is standing on one leg when the rain stopped. It was time to continue with the circular walk, seeing Lough Furnace from all sides and enjoying the distant views of the remarkable pointy shape of Croagh

Patrick. Something I wasn't able to sense, however, was the fight for life being conducted below the glittering surface of this lough.

The salt water in the deep basin of Lough Furnace is held in place by its high density. It is replaced on rare occasions, called renewal events, when water of even greater density arrives in the tidal channel connecting the lough to the sea. Although the details

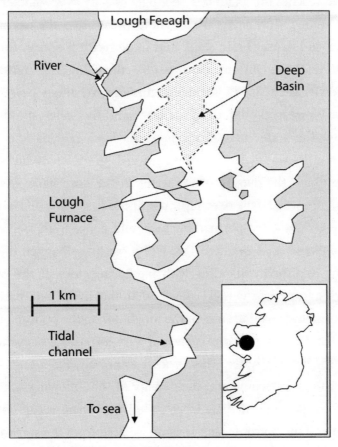

Lough Furnace.

263

may change, the basics of the renewal process are much the same for any deep, isolated basin. At high tide, salt water from the sea creeps towards the loch along the bottom of the tidal channel under a current of fresh water moving seawards (in Lough Furnace, this bottom inflow only happens during a high spring tide). The bottom flow mixes with the fresh water in the tidal channel and by the time it reaches the lough it is no longer as salty as when it left the sea. As the inflow enters the lough, it spreads over the top of the saltier and denser basin water and underneath the fresh surface layer, creating an intermediate layer like the filling in a sandwich.

Slowly, very slowly, turbulent mixing exchanges parcels of water between the intermediate layer and the water above and below. The water that mixes upwards is carried back out to sea, but the water that mixes downwards gradually reduces the salinity and the density of the water in the deep basin. As the mixing proceeds, the deep layer continues to get less dense until a day arrives when particularly salty and dense water comes in through the tidal channel. This will most likely happen during a long dry spell coinciding with big spring tides. If the water arriving from the channel is denser than that in the deep basin, it will tumble down the sides of the lough, filling the bottom of the basin and creating a clean, new and – importantly – oxygenated layer of bottom water. This process continues for a few days. Then, the rains come and the density of the inflowing water is reduced; the intermediate layer once more forms a cap on the basin and the bottom water is again isolated from the surface. In Lough Furnace, it can be several years between renewal events.

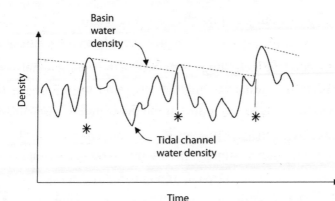

Density

Basin
water
density

Tidal channel
water density

Time

*Renewal events in a deep basin. The stars mark the occasions
when the basin water is replaced.*

The sketch above might help you to picture this process. The density of the water in the tidal channel fluctuates with the state of the tide and weather conditions. Renewal events happen when the water in the tidal channel is denser than that in the deep basin. A renewal event quickly increases the density of the water in the basin, and in the months and years that follow, mixing very gradually reduces it until the time arrives for the next renewal.

The slow mixing of the deep water in the lough is an important part of the renewal process. Without it, the density of the water in the bottom of the basin would stay constant at the highest level that had occurred up to that point. Renewals would only happen when the water emerging from the inland end of the tidal channel was *denser than it had ever been before.* Over time, renewal events would be harder to achieve and they would become less and less frequent. A similar thing happens, I think, with high jump world records: they are broken only when some super-human comes along who can jump higher than anyone before. With the slow mixing of the deep basin, however, renewals of the basin water become inevitable, albeit rare, events. The mixing continually reduces the density of the water in the basin until a day comes when it is lighter than the water in the tidal channel. To continue the high jump analogy, the bar is steadily being lowered until someone can jump it: super-humans are no longer required.

There is a price to be paid, though, for the renewal events. Mixing the lough requires energy. This is because light water will naturally sit on top of heavy water; pushing intermediate-layer water downwards into the dense deep layer requires work to be done against buoyancy. Observations of the changes in salinity of the deep water can be used to calculate exactly how much energy is used in mixing. The answer is surprisingly small; the power requirement for the deep mixing of the whole lough is just 4 watts. That's about the power consumption of a single energy-saving light bulb. I think even I could perform the mixing at that rate for a short while, given a long enough spoon. A likely source of this energy is the internal seiches rocking the water back and forth in the loch: ultimately the energy is taken out of the wind that sets up these seiches.

Looking back on the last few paragraphs, I find it surprising that so many words are needed to explain the working of the small and apparently peaceful body of water that is Lough Furnace. The thing that makes the lough special is that it is almost a fresh-water lake, but not quite. Salt water is able to travel from the sea along the tidal channel and fill the deep basin in the lough, but the journey is a difficult one. There are virtually no tides in the lough and the wind doesn't seem able to get a firm enough grip to give it a good stir. It is the interaction between the salt and fresh water, in conditions where opportunities for mixing are so limited, that makes this water body so fascinating.

It is not far to drive from Lough Furnace along the north shore of Clew Bay to Achill Head, which sticks out from the west coast of Ireland like an index finger pointing towards the Atlantic Ocean. From the car park at the nearby beach, a very steep, rough path leads up the headland to an observing station, built during the Great War as a lookout point for submarines. This spot was chosen because it gives superlative views across the sea to the south, west and north (I'm often impressed when finding old military establishments at just how good the soldiers were at choosing the best place to put things). At this altitude, the distance to the horizon is about 48km (30 miles). The continental shelf is narrow here, less than 32km (20 miles) from the shore to the shelf edge. Standing on the headland and looking out to sea, continental shelf water is crashing on the rocks below but further out, just below the horizon, is ocean water several kilometres deep. This is one of the few places in these islands where you can look at water that is truly part of the Atlantic Ocean.

The space up here is big. The next land in front of you is Labrador, 2,000 miles to the west. The basin of the North Atlantic is a thousand times wider and 200 times deeper than Lough Furnace but, remarkably, the processes happening in the tiny lough can teach us something about the way this great ocean works. The water of all the oceans takes part in a global circulation that pumps heat from the Equator to the poles much like the way the central heating system in a house distributes the heat from the boiler. The 'boiler' of the Earth is the belt around the Equator. Here, the surface waters of the ocean are warmed by the sun and they flow (largely in surface wind-driven currents such as the North Atlantic Drift) towards high latitudes, giving up their heat to the atmosphere on the way. By the time the water has reached the polar seas, it is cold and dense enough to sink down into the deepest parts of the ocean. The cold water then spreads out along the ocean floor, travelling back towards the Equator. On its way, it rises, slowly, back to the surface to complete the circulation.

The water sinking at the poles brings oxygen to the deep basins of the oceans and allows air-breathing animals to live there. The water can only sink right to the ocean bed, however, if it is *denser* than the water that is already there. The sinking is made possible (just like in Lough Furnace) by the slow mixing of the deep waters of the ocean with the warmer water above. The mixing reduces the density of the bottom water and allows it to make way for new dense water sinking at the poles. In the deep ocean (again just like Lough Furnace), the mixing between layers is thought to be achieved by turbulence created by internal waves on the various density layers. The waves may be internal seiches created by the wind as they are in Lough Furnace; these ocean seiches, however, will have very long wavelengths, stretching between continents. Tides will also make internal waves as they squeeze the layers of the ocean over the top of a submarine mountain, such as the mid-ocean ridges. On the lee side of the mountain, the layers will expand again and as they do so, they oscillate up and down as they try to find the right level for their density. Energy, from wind and tides, is needed to keep the deep circulation of the ocean moving.

I'm sure there are times when the weather is fine on Achill Head, but the day of my visit is not one of them. The wind is at gale force off the sea and it is not easy to stand upright. A couple of young visitors by the observing station have their arms outstretched and their bodies angled to the vertical as the force of the wind supports their weight. White horses abound on the surface of the water and there are no boats in sight. Today, the sea is showing its teeth. It is a reminder that the ocean can be a dangerous place, not the natural element for weak humans and their puny ships. This would be a bad day to study the ocean at close hand.

It *is* possible to learn how the ocean works without going to sea. Automated instruments can be placed on unmanned submarines. Satellites routinely make measurements of the surface of the entire ocean. Samples can be collected for you by merchant vessels. Computers can be programmed to simulate the ocean and predict its future. But I feel sorry for the investigators who never leave their laboratory or computer screen. They are missing an opportunity. The ocean is the last great unexplored frontier on our planet. The scientific study of the ocean is a subject for people with an adventurous spirit. Nothing compares with being at sea, looking out on a wide horizon and coming face to face with this often hostile – but also beautiful – environment that we seek to understand.

ARE WE NEARLY THERE?

O NE OF THE GREAT THINGS ABOUT living in these islands is that nowhere is very far from the sea. According to the Ordnance Survey, the furthest point from salt water on the British mainland is in Derbyshire, near the village of Coton-in-the-Elms. This spot is an equal distance of 70 miles (112.6km) from three places on the coast, in North Wales, Lincolnshire and Gloucestershire. It is a piece of countryside much like any other. There is no official marker at the location, but then it would be wrong-headed if there were. Our seas are tidal and the tide drives salt water in and out of rivers, sometimes over a distance of several miles. If you define the coast as where you first encounter salt water, then the place that is furthest from the coast will move around with the tide. The people at the Ordnance Survey have thought of this and they calculate distances from the low tide mark. But since the time of low tide is not the same in North Wales, Lincolnshire and Gloucestershire, it is not clear that the point furthest from the sea will ever be the one the OS have chosen. As the tide rises and falls in its own time around the coast, the point on the British mainland that lies at the greatest distance from the sea moves in a convoluted circuit (which no one, as far as I know, has ever attempted to draw) around the Midlands of England.

Setting aside the difficulties of a tidally flexible coastline, it can be a fun thing to find the point on an irregularly shaped island that is the most remote from the sea. I don't know how the Ordnance Survey did it, but it seems to me that one way you

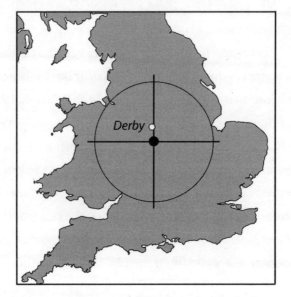

might set about solving the problem is to find the largest circle that you can fit into a map of the country, jiggling it around until the circumference just touches the coastline at two or three points (whenever I tried this with real islands it was always three points, but I think it might be possible with just two if the island has the right shape). The centre of this circle will then be the furthest place from the sea. Applying this method to a map of the British mainland puts the centre of the circle in Derbyshire, although I'm not sure I could pinpoint it as exactly as Coton-in-the-Elms. The centre is equidistant from the Rivers Dee and Severn and the Wash. When you do this exercise, you realise how much the answer you get depends on how you draw the coastline, particularly how far you allow salt water to intrude up long rivers like the Severn and the Trent.

Repeating the circle-fitting exercise for Scotland, I reckon the furthest point from the sea is in the Cairngorms, between Kingussie and Pitlochry, at equal distances of 43 miles (69.2km) from the coast at Loch Linnhe, the Firth of Tay and the Moray Firth. The largest circle I can fit into Wales touches the coast at just two places – the Dovey estuary and the Severn, close to the border at Chepstow – but the other consideration here is getting the centre of the circle inside the Welsh border. The point in Wales furthest from the coast is in the Radnor Forest, right on the border with England and 42 miles (67.5km) from the sea. There happens to be a small Stone Age monument near here called the Four Stones. Faith and I went to see this and wondered, like many other visitors, I imagine, about the people who put up these stones and their reason for doing so. The stones took a bit of finding. A group of locals chatting to the bin men helped; they gave us directions in an accent that would excite the casting director of *The Archers*. When you get to the site, there isn't much to see: just four large stones, one of which has fallen over, on the edge of a field. But there is something about the place. I wonder if we are the first people to visit this site knowing that this is just about as far as you can get from sea water in Wales.

In the case of the island of Ireland, something interesting happens when you try the circle-fitting experiment. There are *two* possible circles, of nearly equal size, one fitting in the northern half of the island and the other in the south. This double solution is produced by the hour-glass shape of an island that is squeezed in the middle by Galway Bay. This gives two contenders for furthest point from the sea, separated by some distance. The southernmost circle is, I think, slightly larger and this has a centre in the Slieve mountains, 60 miles (96.5km) from the sea in three directions: towards Galway Bay, Tramore on the south coast and Wicklow in the east.

ACKNOWLEDGEMENTS

MY THANKS GO TO EVERYONE WHO made this book possible. Jonathan Eyers of Bloomsbury Publishing saw the potential in the early idea and convinced his colleagues to give it a try. He has been a great support on every step of this journey. Clara Jump helped me to put the final copy of the book together. Martyn Roberts and Jenni Davis read early versions of the text and made valuable suggestions for improvement. Sarah Warden drew the wonderful illustrations that adorn several pages of the book. I only wish there could be more of them.

Many of the scientific stories are not mine but belong to colleagues and I am grateful to them for sharing their ideas with me. They include Paul Butler of the University of Exeter, Hannah Byrne, Caren Binding of the Canada Centre for Inland Waters, Alan Elliott, Francis Gohin of IFREMER in Brest, Kevin Horsburgh of the National Oceanography Centre, John Hunter, Jonathan Malarkey of Bangor University, Gay Mitchelson-Jacob, Rick Nunes-Vaz, Jonathan Sharples of Liverpool University,

David Thomas of Helsinki University, the late Marcel Wernand of the Royal Netherlands Institute for Sea Research, Martin White of the National University of Ireland in Galway and Phil Woodworth of the National Oceanography Centre in Liverpool. Phil Woodworth was also a great help with other aspects of the book, including showing me how I might access the archives of the Liverpool Observatory and Tidal Institute. Most of all from a science perspective, though, my thanks go to John Simpson. It was he who taught me, as he did so many, that shelf seas are fascinating and important.

Rachel Ashe, the founder and director of Mental Health Swims, told me about the benefits of sea swimming. Tony Scriven, a fisherman from Southwold, recalled his time as a volunteer observer of sea temperatures. Others helped with the search for information. They are John Williams and Kevin Hodge of the United Kingdom Hydrographic Office; Marc Duggan of Bangor University Library; Wendy Simkiss of Liverpool Museums; Olga Andres of the Centre for Environment, Fisheries and Aquaculture in Lowestoft; Carol Giles and Tamar Atkinson of the Marine Biological Association on Citadel Hill in Plymouth. In several cases, I think, these people went beyond the usual call of duty and I am very grateful for that.

I tried to visit the sites, some of them at least, connected with the discoveries recounted here. These journeys were much more fun and less traumatic when my wife came with me – she stopped me leaving a trail of personal possessions behind. In fact, Faith supported me in so many ways while I was writing the book, as she has done, quietly and competently, throughout our married life. I couldn't have done this without her.

Anyone who would like more scientific detail about the stories in this book could do no better than consult the *Introduction to the Physical and Biological Oceanography of Shelf Seas* by John H Simpson and Jonathan Sharples. Indeed, I opened and read this work many times when writing this book.

PICTURE CREDITS

THE PICTURES OF HMS *CYCLOPS* AND the Edgell Building in Chapter 1 were kindly provided by the United Kingdom Hydrographic Office. The drawing of the helical diatom chain in Chapter 3 is by Sarah Warden and is based on an original micro-photograph of a drop of sea water taken by Paul Smith. Permission to reproduce the satellite infrared image in Chapter 3 has been granted by NEODAAS. The original image was produced by the University of Dundee Satellite Receiving Station. The circulation map in Chapter 3 was redrawn from an illustration by AJ Elliott and T Clarke, published in *Continental Shelf Research* in 1991. The drawing of the Alma Inn in Chapter 5 is by Sarah Warden and is based on a photograph of the inn on the 'Historic Harwich Pub Trail' website. The drawing of the *Salpa* in Chapter 9 is also by Sarah Warden and is based on pictures supplied by the Marine Biological Association of the United Kingdom. The map of surface salinity in the Clyde

Sea in Chapter 9 is redrawn from a picture in an article by CE Binding and myself, published in *Estuarine, Coastal and Shelf Science* in 2003. Sarah Warden drew the lighthouse pictures in Chapters 2 and 4 and the shell in Chapter 3.